Charles W. Groetsch

Inverse Problems in the Mathematical Sciences

Charles W. Groetsch

Inverse Problems in the Mathematical Sciences

With 38 Illustrations

Die Deutsche Bibliothek – CIP-Einheitsaufnahme

Groetsch, Charles W.:
Inverse problems in the mathematical sciences /
Charles W. Groetsch. – Braunschweig; Wiesbaden:
Vieweg, 1993
ISBN 3-528-06545-1

Professor Charles W. Groetsch
Department of Mathematical Sciences
University of Cincinnati
Cincinnati, Ohio 45221-0025
USA

Mathematical Subject Classification: 00A69, 45B05, 65R30, 45B05

Originally published by Friedr. Vieweg & Sohn Verlagsgesellschaft mbH, Braunschweig/Wiesbaden, in 1993.

Vieweg is a subsidiary company of the Bertelsmann Publishing Group International.

Cover design: Klaus Birk, Wiesbaden

Printed on acid-free paper

ISBN 978-3-322-99204-8 ISBN 978-3-322-99202-4 (eBook)
DOI 10.1007/978-3-322-99202-4

Preface

Classical applied mathematics is dominated by the Laplacian paradigm of known causes evolving continuously into uniquely determined effects. The classical *direct* problem is then to find the unique effect of a given cause by using the appropriate law of evolution. It is therefore no surprise that traditional teaching in mathematics and the natural sciences emphasizes the point of view that problems have a solution, this solution is unique, and the solution is insensitive to small changes in the problem. Such problems are called *well-posed* and they typically arise from the so-called direct problems of natural science. The demands of science and technology have recently brought to the fore many problems that are *inverse* to the classical direct problems, that is, problems which may be interpreted as finding the cause of a given effect or finding the law of evolution given the cause and effect. Included among such problems are many questions of remote sensing or indirect measurement such as the determination of internal characteristics of an inaccessible region from measurements on its boundary, the determination of system parameters from input-output measurements, and the reconstruction of past events from measurements of the present state. Inverse problems of this type are often *ill-posed* in the sense that distinct causes can account for the same effect and small changes in a perceived effect can correspond to very large changes in a given cause. Very frequently such inverse problems are modeled by integral equations of the first kind.

The level of research activity in integral equations of the first kind, inverse problems and ill-posed problems have been very high in recent years, however, the rank-and-file teaching faculty in undergraduate institutions is largely unaware of this exciting and important area of research. This is a double tragedy because of the scientific importance of the topic and the fact that many of the concepts and ideas involved in the study of integral equations, inverse problems and ill-posed problems can be introduced in the undergraduate curriculum.

This monograph arose from a National Science Foundation Faculty Enhancement Project. The goal of the project was to apprise college faculty of some of the main lines of research in inverse and ill-posed problems and to present some models and methods at a fundamental level. The monograph is therefore a broad based introduction rather than a comprehensive survey of the field. I would like to thank my co-conspirators in the NSF project, Zuhair Nashed, Gil Strang, Milt Wing and John Zahrt for helping to make the project possible. Special thanks are due to my TEX-nician, Anne Feldman and TEX-nical advisor, Chris McCord.

C.W. Groetsch

Contents

1 Introduction

Is there a thing of which it is said, "See, this is new"? It has been already in the ages before us.
Ecclesiastes 1:10

The study of inverse problems is very new — and very old. The latest high-tech medical imaging devices are essentially inverse problem solvers; they reconstruct two or three-dimensional objects from projections. More than two thousand years ago, in book VII of his *Republic*, Plato posed essentially the same problem in his allegory of the cave, namely, he considered the philosophical implications of reconstructing "reality" from observations of shadows cast upon a wall.

These notes are meant to provide a first look at inverse problems in the mathematical sciences. It would be nice at this point to be able to give a clear, crisp definition of an inverse problem in the same way that one defines "group" or "topological space". However, the situation is not quite so clear-cut for inverse problems because the posing of an inverse problem presupposes the statement of a direct problem. In trying to give a general definition of an inverse problem we find ourselves in a position akin to that experienced by Justice Potter Stewart who, in referring to pornography, said he couldn't define it, but he knew it when he saw it.

An *inverse problem* is a problem which is posed in a way that is inverted from that in which most *direct* problems are posed. The type of direct problem we have in mind is that of determining the effect y of a given cause x when a definite mathematical model K is posited: $Kx = y$. For such direct problems we assume that the operator K is well-defined and continuous, therefore there is a unique effect y for each cause x and small changes in x result in small changes in y. But this direct problem is only one third of the story. Given a direct problem of the type just discussed, two inverse problems may be immediately posed. These are the inverse problems of causation (given K and y, determine x) and model identification (given x and y, determine K). In the direct problem existence, uniqueness and stability of solutions is assumed, but in inverse problems none of these qualities can be taken for granted and it is this that makes inverse problems challenging and mathematically interesting.

Inverse problems in the physical sciences have been posed throughout the historical development of the subject as a mathematical discipline. Corresponding to the direct problem of determining the resistive force on a solid of revolution of specified shape moving through a fluid, Newton proposed the inverse problem of determining a shape giving rise to a given resistive force. Similarly, Huygens in his design of an isochronous pendulum clock, and Bernoulli in his study of paths leading to a given time of descent, studied problems which are inverse to the standard direct problem of time of descent on a given curve. The inverse problems just mentioned had a profound influence on mathematics and led to the founding of a new field of study – the *calculus of variations.* Inverse problems have also led to major physical advances, perhaps the most spectacular of which was the discovery of the planet Neptune after predictions made by Leverrier and Adams on the basis of inverse perturbation theory. In his 1848 book on the work of Leverrier and Adams, J.P. Nichol took a surprisingly modern, though overly optimistic, view of inverse theory relating to the figure of the Earth as deduced from the orbit of the moon: "Certain deviations are caused by the influence of our equatorial protuberance: and these deviations – measured by our modern instruments, whose precision approaches to the marvelous – enables us, by inverse reasoning, to determine with undoubted exactness, how far the Earth deviates from a regular globe."

A common feature of inverse problems posed in function spaces is their *instability*, that is, small changes in the data may give rise to large changes in the solution. The computational treatment of such problems requires some type of discretization to fashion an approximate problem in which there are only finitely many unknowns. Small finite dimensional problems are typically stable, however, as the discretization is refined to better model the original infinite dimensional problem, the number of variables increases and the instability of the original problem becomes apparent in the discrete model. Nichol was evidently unaware of the difficulty of instability in inverse problems, but other authors of the last century were remarkably prescient of the issue of instability. Maxwell noted in 1873, "There are certain classes of phenomena ... in which a small error in the data introduces a small error in the result ... The course of events in these cases is stable. There are other classes of phenomena which are more complicated and in which cases of instability occur, the number of such cases increasing, in an extremely rapid manner, as the number of variables increases."

Around the turn of the century, Hadamard clearly formulated the concept of a *well-posed* problem. He took existence, uniqueness and stability of solutions to be the characteristics of a well-posed problem and expressed the view that physical situations always lead to well-posed problems. This view was echoed by Petrovskii as lately as 1956 and we find in Courant and Hilbert the statement "... a mathematical problem cannot be considered as realistically corresponding to physical phenomena unless a variation of the given data in a sufficiently small range leads to an arbitrarily small change in the solution." Interestingly, only three pages on from this quotation, we find " 'properly posed' problems are by far not the only ones which appropriately reflect real phenomena." Today we know that many interesting and important inverse problems in science lead to mathematical problems that are not well-posed in the sense of Hadamard. The major goal of this work is to introduce simple examples of such problems, consider the challenges they present, and introduce the basics of some methods designed to meet those challenges.

The many important inverse problems arising in modern science and technology more than justify a greater role for inverse theory in the undergraduate curriculum. But more important than scientific applications is the need to teach students the value of "inverse thinking", irrespective of the particular field of investigation. Such thinking can lead to fresh perspectives and audacious ideas. We conclude this introduction with two examples of this type of thinking in fields far removed from mathematics.

In studying insect populations, say moths, one is led directly to consider the food source of the population. Charles Darwin, in 1862, turned this thinking around by predicting the existence, on Madagascar, of an unknown moth with an eleven inch proboscis. His prediction was based on the existence of a Madagascan orchid with a foot-long nectary. Some forty years later such a moth was discovered on Madagascar, dramatically confirming Darwin's prediction. The analogy with the discovery of Neptune is inescapable.

Our final example comes from biblical archaeology. The story of the discovery of the Dead Sea scrolls, beginning in 1947, is now familiar. What is not so well-known is that a small group of scholars was given, by the Jordanian Department of Antiquities (with the acquiescence of the Israeli authorities after the 1967 war), exclusive rights to the source materials. The so-called International Committee routinely denied other scholars access to the materials while publishing its own results at an excruciatingly slow pace. The committee did, however, publish a concordance of the scrolls in the 1950's. This concordance (essentially a collection of "literary projections" of the scrolls) was readily available to all scholars. Two of the scholars, Ben-Zion Wacholder and his student Martin Ebegg, in their frustration at being denied access to the scrolls, conceived the bold idea of reconstructing the scrolls from the concordance – a kind of *literary tomography*. Using a computer they did just that. The result, though far from accurate, introduced enough instability into the political situation to break the monopoly on the scrolls and allow access by all scholars to the 2,000 year old mother lode. Inverse problems are very new and very old.

2 Inverse Problems Modeled by Integral Equations of the First Kind: Causation

Pangloss could prove to everybody's satisfaction
that there is no effect without a cause
Voltaire, Candide

What causes a given effect? Often this question has no definitive answer. We may be able to suggest a number of distinct possible causes for an effect and sometimes we can find no reasonable cause at all that can account for a given effect (such an effect we call a *mystery*). Most disconcertingly, it may happen that highly disparate causes result in indistinguishable effects.

In this chapter we investigate causation in a number of concrete physical situations. A cause will be a function $x = x(t)$ and the effect $y = y(s)$ of this cause is felt through a deterministic process $K : y = Kx$. Schematically, the cause-effect relationship is illustrated in Figure 2.1.

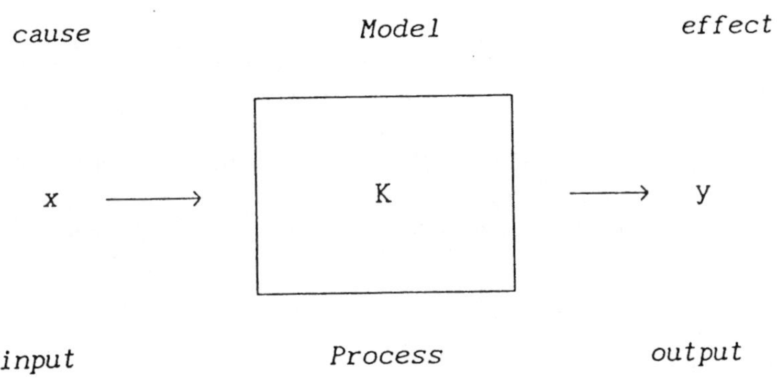

Figure 2.1

Very often the process, or our idealization of it, is *linear*, that is,

$$K(c_1x_1 + c_2x_2) = c_1Kx_1 + c_2Kx_2.$$

Moreover, the details of a given effect often result from a combination of the details of the cause. A mathematical leap then leads us to the following form for the cause-effect relationship:

$$y(s) = \int_a^b k(s,t)x(t)dt \tag{2.1}$$

where the function $k(\cdot,\cdot)$ represents a (simplified) model of the process. If the linearity assumption is dropped, the process has the more general form

$$y(s) = \int_a^b k(s,t,x(t))dt$$

where $k(\cdot,\cdot,\cdot)$ is a given function of three variables. We shall concentrate on equations of the form (2.1), which are called *Fredholm integral equations of the first kind* ($k(\cdot,\cdot)$ is called the *kernel* of the equation).

When the variable t represents time and the past is unaffected by the future, then $k(s,t) = 0$ for $s < t$ and (2.1) takes the form

$$y(s) = \int_a^s k(s,t)x(t)dt. \tag{2.2}$$

This special form of (2.1) is called a *Volterra integral equation of the first kind* and we shall see that such equations model many temporal and nontemporal physical situations. Sometimes the kernel in (2.2) exhibits a special translational invariance because it depends only on the difference of the arguments, that is, $k(s,t) = k(s-t)$. In this case (2.2) becomes

$$y(s) = \int_a^s k(s-t)x(t)dt \tag{2.3}$$

which is called a *convolution* equation. The Laplace transform is a standard tool for the analysis of convolution equations.

Each of the four types of integral equations of the first kind introduced above will occur among the models in the next section.

2.1 Some Models

We now present a number of inverse problems in physical science arising in statics, dynamics, potential theory, heat transfer, hydraulics, imaging, radiation, diffusion and biochemical reactions. In each case the inverse problem will be modeled in terms of an integral equation of the first kind. For the most part, the models considered are *linear* and they therefore represent, as do all models, a simplified representation of physical reality. We begin by taking up a problem in statics.

The Hanging Cable. Imagine a cable of variable density hanging between two horizontal supports. We assume that the tension T in the cable is constant and that the vertical deflection y of the cable at any point is small relative to the length of the cable. A somewhat exaggerated (remember, we assume small vertical deflections) illustration of the situation is:

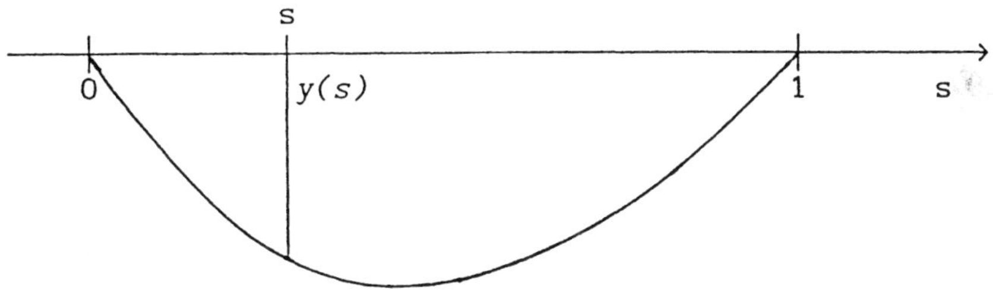

Figure 2.2

The inverse problem we wish to pose is this: what distribution of the variable mass of the cable causes the observed deflection mode y? We will call the weight density of the cable $x(s)$ and we construct a model K for the weight-deflection relationship: $y = Kx$.

Consider the effect of a concentrated force F at the point t (see Figure 2.3). Balancing forces we find:

$$T sin\phi + T sin\theta = F.$$

Because of the small deflection assumption, we have $\sin\theta \approx \tan\theta$ and $\sin\phi \approx \tan\phi$ and hence we model the balance of forces by

$$\frac{y(t)}{t} + \frac{y(t)}{1-t} = \frac{F}{T}$$

and therefore

$$y(t) = \frac{F}{T}t(1-t).$$

If $s < t$, then by similarity of the triangles

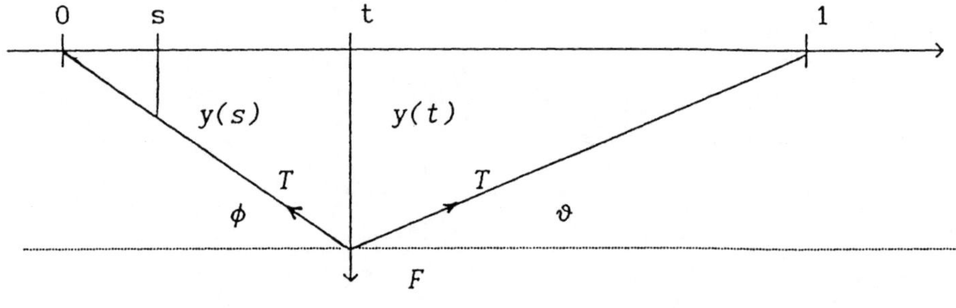

Figure 2.3

$$\frac{y(s)}{s} = \frac{y(t)}{t} = \frac{F}{T}(1-t)$$

or

$$y(s) = \frac{F}{T}s(1-t).$$

Similarly,

$$y(s) = \frac{F}{T}t(1-s) \text{ for } s > t.$$

Hence we have

$$y(s) = F\,k(s,t)$$

where

$$k(s,t) = \begin{cases} t(1-s)/T, & 0 \le t \le s \\ s(1-t)/T, & s \le t \le 1. \end{cases}$$

Consider now a continuous distribution of force produced by a weight density $x = x(t)$. Taking into account the influence of all the infinitesimal forces on the deflection $y(s)$ we arrive at

$$y(s) = \lim_{n\to\infty} \sum_{i=1}^{n} k(s,t_i)x(t_i)\Delta t_i$$

or

$$y(s) = \int_0^1 k(s,t)x(t)dt \tag{2.4}$$

which is a Fredholm integral equation of the first kind relating the density to the deflection.

Exercise 2.1: Show that if y satisfies (2.4), then y is a solution of the boundary value problem

$$\begin{aligned} y''(s) + x(s) &= 0, \quad 0 < s < 1 \\ y(0) = y(1) &= 0. \end{aligned}$$

Show that a small perturbation $y_\epsilon(s) = \epsilon(s-1)\sin(s/\epsilon)(\epsilon \ll 1)$ in the deflection y is accounted for by a large perturbation in x. □

Exercise 2.2: Consider a modification of the hanging cable model, namely the problem of a shaft rotating with angular velocity ω in which loads result from centrifugal forces generated by small deflections of the center line of the shaft. Show that the deflection of the center line $y(s)$ is related to the weight density $x(t)$ by

$$y(s) = \omega^2 \int_0^1 k(s,t)y(t)x(t)dt. \qquad \square$$

Geological Prospecting. The general problem of geological prospecting is to determine the location, shape and constitution of subterranean bodies from measurements at the earth's surface. We will treat a much simplified one-dimensional model. Namely, we suppose that on a horizontal line measurements are made of the vertical component of gravitational force due to a variable distribution of mass along a parallel line one unit below the surface. The situation is illustrated in Figure 2.4.

If the variable mass density $x(t)$ is distributed along the t-axis for $0 \le t \le 1$, and an instrument measures the vertical component of force $y(s)$, then a small mass element $x(t)\Delta t$ at position t gives rise to a vertical force $\Delta y(s)$ at s given by

$$\Delta y(s) = \gamma \frac{x(t)\Delta t}{(s-t)^2+1}\cos\theta = \gamma \frac{x(t)\Delta t}{((s-t)^2+1)^{3/2}}$$

where γ is the gravitational constant. Therefore, taking into account all such elements Δt, we find that the vertical force $y(s)$ at s is related to the density distribution $x(t)$ by way of the following Fredholm integral equation of the first kind:

$$y(s) = \gamma \int_0^1 ((s-t)^2+1)^{-3/2} x(t)dt. \tag{2.5}$$

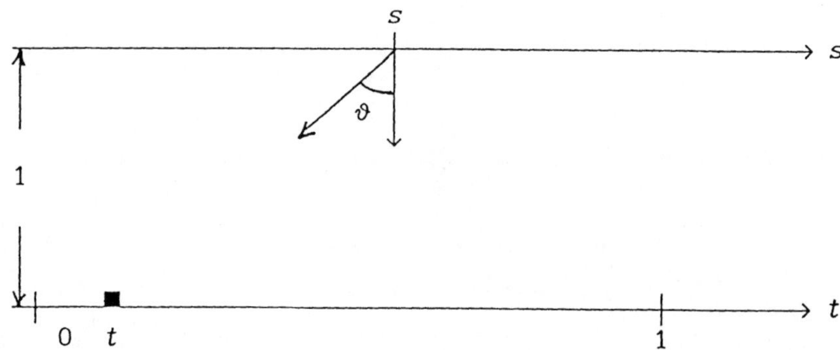

Figure 2.4

Exercise 2.3: Show that (2.5) has at most one solution. (Hint: Extend $x(t)$ to $(-\infty, \infty)$ by setting $x(t) = 0$ for $t \notin (0,1)$ and apply the Fourier transform.) □

Exercise 2.4: Show that the centrally directed force $y(\theta)$ at a point θ on the unit circle due to a variable mass density $x(\varphi)$ distributed over a concentric circle of radius 1/2 is given by

$$y(\theta) = \gamma \int_0^{2\pi} \frac{1 - 1/2\cos(\theta - \varphi)}{(5/4 - \cos(\theta - \varphi))^{3/2}} x(\varphi) d\varphi. \qquad \square$$

Exercise 2.5: Consider a two-dimensional version of the gravitation model in which a body Ω contained within the unit disk has variable density $x(\vec{r})$. Show that the centrally directed force $y(\vec{R})$ at a point $\vec{R}$ on the unit circle is given by

$$\begin{aligned} y(\vec{R}) &= \gamma \int_\Omega \frac{(\vec{R} - \vec{r}) \cdot \vec{R}}{\mid \vec{R} - \vec{r} \mid^3} x(\vec{r}) dA(\vec{r}) \\ &= \gamma \vec{R} \cdot \vec{\nabla} \varphi(\vec{R}) \end{aligned}$$

where dA is an area element and φ is the gravitational potential:

$$\varphi(\vec{R}) = \int_\Omega \frac{1}{\mid \vec{r} - \vec{R} \mid} x(\vec{r}) dA(\vec{r}). \qquad \square$$

Exercise 2.6: Use Green's theorem to show that if f is a C^2 function which vanishes along with its normal derivative on the boundary of Ω, then

$$\int_\Omega \frac{1}{|\vec{r} - \vec{R}|} \Delta f(\vec{r}) dA(\vec{r}) = 0$$

where Δf is the Laplacian of f. Conclude that the integral equation in Exercise 2.5 does not have a unique solution. □

The Edge Effect. In this "flat earth" model we assume that the surface of the earth is a horizontal plane. Depth will be measured along a positive z-axis, pointing down, and we assume a stratified model in which the density $\rho(z)$ is a function of z alone. We assume that the z-axis represents an edge between two laterally uniform structures with densities ρ_1 and ρ_2, respectively, such as might exist at a continental margin:

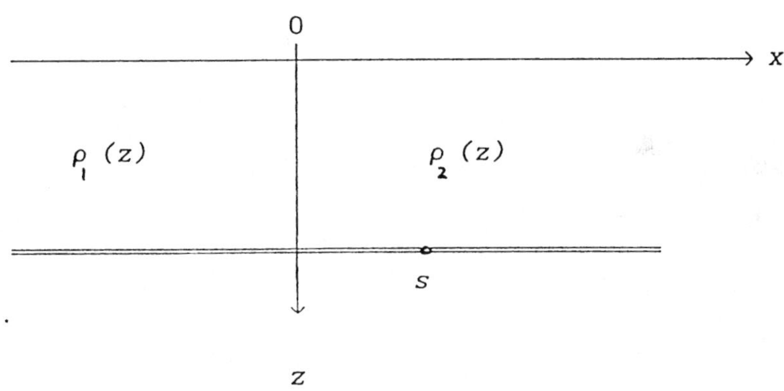

Figure 2.5

If we imagine a y-axis coming out of the page, then it is not difficult to compute the vertical component of gravity $g(x, s, z)$ at the point $(x, 0, 0)$ on the surface engendered by the infinitely long "wire" (s, y, z), $-\infty < y < \infty$, at depth z. In fact

$$\begin{aligned} g(x,s,z) &= \gamma\rho(z,x)z \int_{-\infty}^{\infty} \left\{(x-s)^2 + y^2 + z^2\right\}^{-3/2} dy \\ &= \frac{2\gamma\rho(z,x)z}{(x-s)^2 + z^2} \end{aligned}$$

where

$$\rho(z, x) = \begin{cases} \rho_1(z), & x < 0 \\ \rho_2(z), & x \geq 0. \end{cases}$$

Consider now the gravitational *anomaly* $\Delta g(x, s, z)$ obtained by subtracting out the density $\rho_1(z)$, that is,

$$\Delta g(x, s, z) = \frac{2\gamma\Delta\rho(z)z}{(x-s)^2 + z^2}, \quad s \geq 0$$

where $\Delta\rho(z) = \rho_2(z) - \rho_1(z)$. Integrating over s, we obtain the gravitational anomaly $\Delta g(x, z)$ at x due to a semi-infinite horizontal plate at depth z:

$$\begin{aligned} \Delta g(x, z) &= 2\gamma\Delta\rho(z)z \int_0^\infty \frac{1}{(x-s)^2 + z^2} ds \\ &= 2\gamma\Delta\rho(z)z \left(\tan^{-1}(\frac{x}{2}) + \pi/2\right). \end{aligned}$$

Therefore the gradient of the anomaly, $\Delta g'(x, z) = \frac{\partial}{\partial x}\Delta g(x, z)$, satisfies

$$\Delta g'(x, z) = \frac{2\gamma\Delta\rho(z)z}{x^2 + z^2}.$$

Finally, integrating over all such semi-infinite plates, we obtain the following integral equation relating the gradient of the gravitational anomaly and the density difference:

$$\Delta g'(x) = 2\gamma \int_0^\infty \frac{\Delta\rho(z)z}{x^2 + z^2} dz. \tag{2.6}$$

Exercise 2.7: To simplify notation, write (2.6) as

$$f(x) = \int_0^\infty \frac{\varphi(z)z}{x^2 + z^2} dz.$$

Show that for $\lambda > 0$, $z > 0$,

$$\int_{-\infty}^\infty \frac{1}{x^2 + z^2} e^{-i\lambda x} dx = \frac{\pi}{z} e^{z\lambda}.$$

Conclude that if $\hat{f}$ is the Fourier transform of f, then for $p > 0$,

$$\hat{f}(-p) = \mathcal{L}\{\pi\varphi(z)\}(p)$$

where $\mathcal{L}$ is the Laplace transform. Therefore the integral equation may be formally inverted as follows:

$$\varphi(z) = \frac{1}{\pi}\mathcal{L}^{-1}\{\hat{f}(-p)\}(z). \qquad \square$$

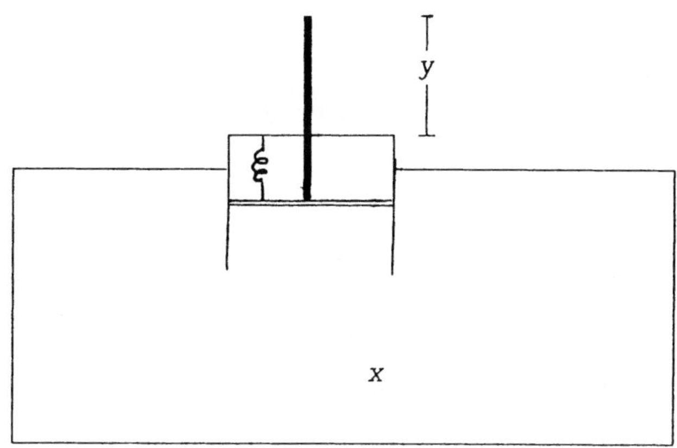

Figure 2.6

Pressure Gauges. Consider a sealed vessel with an attached pressure gauge (Figure 2.6). We suppose that this pressure gauge is a simple spring-loaded piston which for simplicity we assume to be undamped.

Let the mass of the piston be m and the spring constant be k. Suppose that the pressure in the vessel changes in time, due say to heating and cooling of the vessel, and take the cross-sectional area of the piston to be 1, so that force and pressure are equated. The displacement of the gauge above the equilibrium position, $y(t)$, is then related to the internal dynamic pressure, $x(t)$, by

$$my''(t) + ky(t) = x(t).$$

For simplicity we will assume the following initial conditions:

$$y(0) = 0, \qquad y'(0) = 0.$$

Applying the Laplace transform we obtain

$$Y(p) = \frac{1}{mp^2 + k} X(p)$$

and hence, by the convolution theorem,

$$y(s) = \frac{1}{\omega m} \int_0^s \sin \omega(s - t) x(t) dt \tag{2.7}$$

where $\omega = \sqrt{k/m}$. The effect given by the response of the pressure gauge is therefore related to the causing dynamic internal pressure by the convolution equation (2.7).

Exercise 2.8: Derive the integral equation relating the response y to the internal pressure x assuming that the motion is damped by a force proportional to the velocity y'. □

Exercise 2.9: Given an arbitrarily small number $\epsilon > 0$, and an arbitrarily large number $M > 0$, show that there is a pair of functions y_ϵ and x_ϵ satisfying (2.7) with $\max | y_\epsilon(s) | \leq \epsilon$ and $\max | x_\epsilon(t) | \geq M$. □

The Vibrating String. The free vibrations of a nonhomogeneous string of length 1 and density distribution $\rho(x) > 0$, $0 < x < 1$, may be modeled by the partial differential equation

$$\rho(x)u_{tt} = u_{xx}.$$

We take the ends of the string to be fixed and hence the boundary conditions

$$u(0,t) = 0, \qquad u(1,t) = 0$$

are satisfied. The method of separation of variables is a standard technique for analyzing such partial differential equations, that is, a basic solution of the form

$$u(x,t) = y(x)r(t)$$

is assumed. This leads to the ordinary differential equation

$$y'' + \omega^2 \rho(x) y = 0 \tag{2.8}$$

for the spatial component and the boundary conditions

$$y(0) = 0, \qquad y(1) = 0.$$

The numbers ω^2 comprise a discrete sequence of eigenvalues corresponding to allowable frequencies of vibration of the string.

We imagine an observation of y at a given frequency ω, say by employing a stroboscopic light. This observation will be denoted $y(\cdot;\omega)$. From (2.8) we find

$$\int_0^1 y'(s;\omega)ds - y'(0;\omega) + \omega^2 \int_0^1 \int_0^s \rho(x)y(x;\omega)dxds = 0$$

or

$$\int_0^1 sy(s;\omega)\rho(s)ds = \frac{y'(0;\omega)}{\omega^2} \tag{2.9}$$

The inverse problem of determining the variable density ρ of the string then consists of finding a single positive function ρ which satisfies (2.9) for all allowable frequencies ω.

We now take up a couple of inverse problems involving the diffusion of heat.

Thermal Archaeology. Consider a uniform bar of length π (for simplicity) which is insulated on its lateral surface so that heat is constrained to flow in only one direction (the x-direction). With certain normalizations and scalings the temperature $u(x,t)$ satisfies the partial differential equation

$$u_t = u_{xx}, \qquad 0 < x < \pi.$$

We assume the ends of the bar are kept at temperature 0 and that the initial temperature distribution is a function $f(x)$, $0 \leq x \leq \pi$, that is, the boundary and initial conditions

$$u(0,t) = 0, \qquad u(\pi,t) = 0, \qquad u(x,0) = f(x)$$

hold.

A standard direct problem in applied mathematics is to find some subsequent temperature distribution, say at $t = 1$, from this information. If we call this later temperature distribution $g(x) = u(x,1)$, then the situation is illustrated in Figure 2.7.

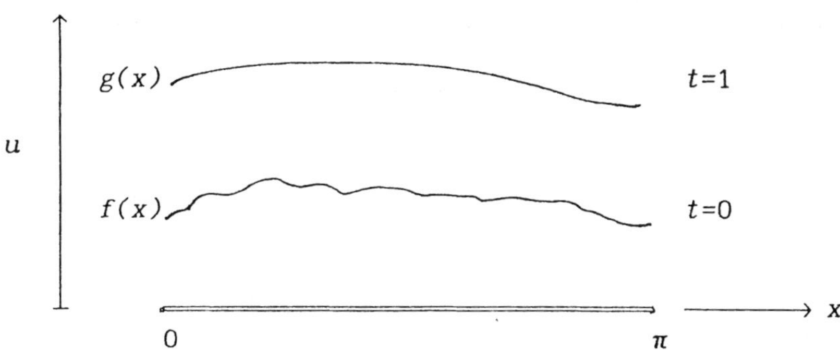

Figure 2.7

The method of separation of variables leads to a representation of $g(x)$ in terms of the eigenfunctions $\sin nx$ of the form

$$g(x) = \sum_{n=1}^{\infty} a_n \sin nx$$

where the coefficients are given by

$$a_n = \frac{2}{\pi} \int_0^{\pi} f(u) \sin nu \ du \, e^{-n^2}.$$

Substituting this into the expression for g and interchanging the summation and integration we arrive at

$$g(x) = \int_0^{\pi} k(x,u) f(u) du \tag{2.10}$$

where

$$k(x,u) = \frac{2}{\pi} \sum_{n=1}^{\infty} e^{-n^2} \sin nx \sin nu.$$

Think now of the *inverse problem*, that is, the problem of determining the initial temperature distribution $f(x)$ that gives rise to the later temperature distribution $g(x)$. This is the problem of solving the heat equation "backward in time." Mathematically it is expressed as solving the integral equation (2.10) for f.

It should be clear from the physical nature of the process that the detailed structure of the initial temperature distribution f is highly diffused at the later time $t = 1$ and hence recovering this detailed information from measurements of g will be exceedingly difficult. The mathematical basis for the difficult reconstruction problem is evident from the form of the kernel in (2.10). Specifically, high frequency components in f (i.e., components associated with $\sin nu$ for large n) are severely damped by the very small factor e^{-n^2}, making their influence on g practically imperceptible. The next exercise makes these ideas concrete.

Exercise 2.10: Suppose f and g satisfy (2.10). Let $\epsilon > 0$ and $M > 0$ be given numbers (ϵ arbitrarily small and M arbitrarily large) and let $f_M(u) = M \sin mu$. Show that an arbitrarily large perturbation f_M in f leads, for m sufficiently large, to a perturbation of maximum amplitude less than ϵ in g. □

Temperature Probes. Suppose a hostile environment is enclosed by a protective wall (the containment vessel of a nuclear reactor is a suitable mental image). It is desired to remotely monitor the internal temperature by passing a long (for our purposes we will assume infinitely long) bar through the wall and measuring the temperature at a point $x = a$ on the safe side of the wall (see Figure 2.8).

If we denote the temperature at the point x on the bar at time t by $u(x,t)$, then the problem is to determine the internal temperature $f(t) = u(0,t)$ from measurements

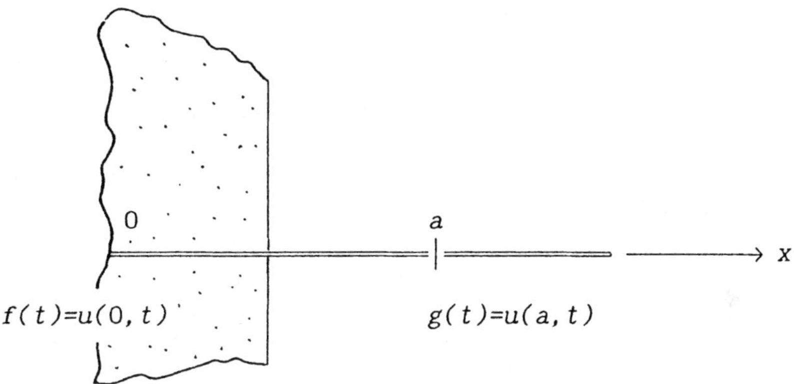

Figure 2.8

of $g(t) = u(a,t)$. We assume that the one-dimensional heat equation is satisfied, that the initial temperature of the bar is 0 and that the temperature is uniformly bounded. Then

$$\begin{aligned} u_t &= u_{xx}, \quad 0 < x < \infty \\ u(x,0) &= 0 \end{aligned}$$

and hence, taking Laplace transforms with respect to the variable t, we arrive at the ordinary differential equation

$$pU = U''$$

where p is the Laplace transform variable, primes signify differentiation with respect to x and U is the Laplace transform of u with respect to t. Bounded solutions of this equation have the form

$$U(x) = A(p)e^{-\sqrt{p}x}$$

where

$$U(0) = F(p),$$

and $F(p)$ is the Laplace transform of $f(t)$. Therefore, by the convolution theorem,

$$u = h * f$$

where h is the inverse Laplace transform of $e^{-\sqrt{p}x}$, that is [AS],

$$h(x,t) = \frac{x}{2\sqrt{\pi}} t^{-3/2} e^{-x^2/4t}.$$

Therefore at $x = a$, we have

$$g(t) = \frac{a}{2\sqrt{\pi}} \int_0^t \frac{\exp(-a^2/4(t-\tau))}{(t-\tau)^{3/2}} f(\tau) d\tau.$$

That is, the internal and external temperatures are related by a convolution equation.

Exercise 2.11: Show that if $u(x,t) = \sqrt{\frac{2}{n}} \sin\left(nt + \sqrt{\frac{n}{2}}x\right)$, then $u_t = u_{xx}$, $u(0,t) = f(t)$, where $f(t) = \sqrt{\frac{2}{n}} \sin nt$ and $u_x(0,t) = g(t)$, where $g(t) = \cos nt + \sin nt$. Show that if $\epsilon > 0$ (arbitrarily small) and $M > 0$ (arbitrarily large), then for any fixed $a > 0$ there are functions f and g satisfying the above conditions with $\max \mid f(t) \mid < \epsilon$, $\max \mid g(t) \mid \leq 2$ and $\max \mid u(a,t) \mid > M$. □

In many, perhaps most, modeling situations the geometrical configuration in which the process acts is assumed to be known. However, for a given physical process changes in geometry can, and usually do, result in changes in effect. For example, the external gravitational field generated by a homogeneous body generally depends upon the shape of the body. In such instances we can think of the geometry as the cause of an observed or given effect. The inverse problem consists of finding this cause, i.e., geometry. The next three examples are very simple illustrations of some inverse problems modeled as integral equations of the first kind in which the solution is some geometrical curve.

Horology. This example is perhaps the oldest instance of an integral equation. The roots of the problem are traced to Huygens and Bernoulli and its formulation as an integral equation is due to Abel in the early nineteenth century. The problem is to find the path in the plane along which a particle will fall, under the influence of gravity alone, so that at each instant the time of fall is a given (or observed) function of the distance fallen.

Suppose, as indicated in Figure 2.9, that the particle falls from height z and that the path of descent is parameterized by arclength s, that is, at time t the length of arc traversed is $s(t)$ ($s(0) = 0$). Assuming that the particle starts from rest, we find by equating the gain in kinetic energy to the loss in potential energy that:

$$\frac{1}{2}\left(\frac{ds}{dt}\right)^2 = g(z-y).$$

Integrating this, we find that the time of descent from z to the base line $y = 0$, $\tau(z)$ is given by

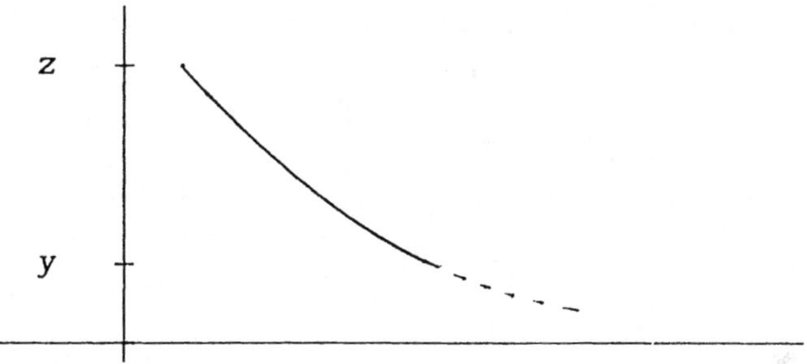

Figure 2.9

$$\tau(z) = \int_{y=z}^{y=0} \frac{ds}{\sqrt{2g(z-y)}}.$$

Setting $\frac{ds}{dy} = -f(y)$, this gives

$$\tau(z) = \int_0^z \frac{f(y)}{\sqrt{2g(z-y)}} dy. \tag{2.11}$$

Exercise 2.12: The *Abel transform* is the integral transform A defined by

$$(A\varphi)(z) = \frac{1}{\sqrt{\pi}} \int_0^z \frac{\varphi(t)}{\sqrt{z-t}} dt.$$

Therefore (2.11) reads: $\tau = \sqrt{\frac{\pi}{2g}} Af$. Show that

$$(A^2\varphi)(x) = \int_0^x \varphi(t) dt.$$

It follows that if D is the differentiation operator, then $DA^2\varphi = \varphi$. Explain why the operator DA may be regarded as "differentiation of one-half order." □

Exercise 2.13: Show that (2.11) can have at most one continuous solution. □

Exercise 2.14: Huygens was interested in an inverse synthesis problem in horology, namely, the design of an isochronic pendulum clock in which the period is independent of the amplitude. In terms of equation (2.11), he wanted to find an f for which $\tau(z) = \tau$ is independent of z. Show that if the time of descent $\tau(z)$ is independent of z, then $(A^2 f)(y) = \alpha\sqrt{z}$, for some constant α and hence

$$\frac{ds}{dy} = -f(y) = \beta/\sqrt{y}$$

for some constant β. Show that this condition is satisfied by the cycloidal arc

$$x = a(\varphi - \sin\varphi), \qquad y = a(1 + \cos\varphi), \qquad 0 \leq \varphi \leq \pi. \qquad \square$$

Irrigation. In traditional agriculture fields are often watered from elevated irrigation canals by removing a solid gate from a weir notch. We suppose that the depth of water in the canal is h and that the notch is symmetric about a vertical center line as in Figure 2.10.

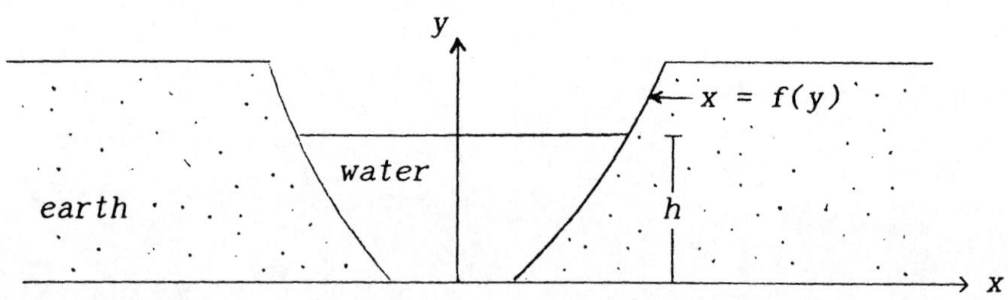

Figure 2.10

By Torricelli's law (see, e.g. Chapter 3), the velocity of the effluent at height y is $\sqrt{2g(h-y)}$, therefore the volume of flow per unit time through the notch is

$$2\int_0^h \sqrt{2g(h-y)} f(y) dy$$

where $x = f(y)$ specifies the shape of the notch. Suppose that one wishes to design a notch so that this quantity is a given function $\varphi(h)$ of the water depth in the canal (or equivalently, suppose one wants to determine the shape f from observations of the flow rate φ). One then is led to solve the convolution equation

$$\varphi(h) = \int_0^h 2\sqrt{2g(h-y)}f(y)dy. \tag{2.12}$$

Exercise 2.15: Show that equation (2.12) has at most one continuous solution. □

Exercise 2.16: Solve equation (2.12) when $\varphi(h) = 2h^2$. □

The Shape of a Mass. We modify a previous example (geological prospecting) a bit in this example. Imagine an airplane in level flight 1 mile high. An instrument on the plane measures the vertical component of the gravitational attraction of a hill of uniform density. The problem is to determine the shape φ of the hill, as pictured in Figure 2.11.

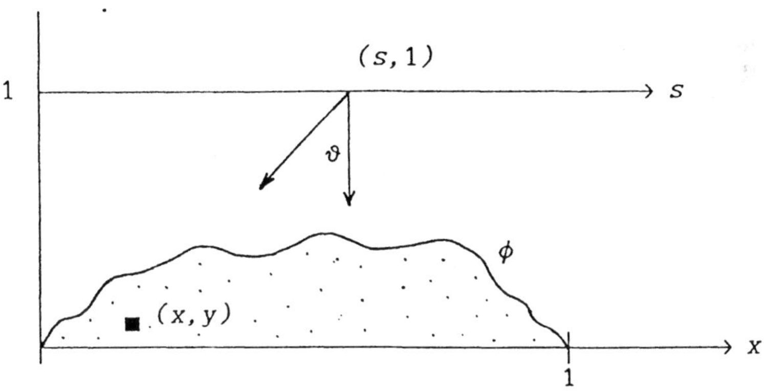

Figure 2.11

The contribution of a small mass element $\rho\Delta x\Delta y$ at the location (x, y) to the vertical component of force at $(s, 1)$ is

$$\frac{\gamma\rho}{(s-x)^2+(1-y)^2}\cos\theta\Delta y\Delta x = \frac{\gamma\rho(1-y)}{((s-x)^2+(1-y)^2)^{3/2}}\Delta y\Delta x$$

where ρ is the density and γ is the gravitational constant. Hence the vertical force $f(s)$ at $(s,1)$ is given by

$$f(s) = \int_0^1\int_0^{\varphi(x)} \frac{\gamma\rho(1-y)}{((s-x)^2+(1-y)^2)^{3/2}}dydx,$$

that is, the shape, φ, of the hill satisfies the *nonlinear* Fredholm integral equation of the first kind

$$f(s) = \gamma\rho\int_0^1 k(s,x,\varphi(x))dx$$

where

$$\begin{aligned} k(s,x,z) &= \int_0^z \frac{(1-y)}{((s-x)^2+(1-y)^2)^{3/2}}dy \\ &= ((s-x)^2+(1-z)^2)^{-1/2} - ((s-x)^2+1)^{-1/2}. \end{aligned}$$

Exercise 2.17: Consider a star shaped body with uniform density contained within the unit circle. Suppose the boundary of the body is described in polar coordinates by $r = g(\theta)$, $0 \le \theta \le 2\pi$. Show that if the centrally directed gravitational force at a point $e^{i\varphi}$ on the unit circle is $f(\varphi)$, then

$$f(\varphi) = \gamma\rho\int_0^{2\pi} k(\varphi,\theta,g(\theta))d\theta$$

where

$$k(\varphi,\theta,z) = \int_0^z \frac{rdr}{(1+r^2-2r\cos(\varphi-\theta))^{3/2}}. \qquad \square$$

We now take up three simple models involving imaging, that is, determining the internal structure of an object from external measurements, or reconstructing a degraded picture.

Simplified Tomography. For simplicity we consider a two-dimensional object (e.g., a simplified tumor) contained within a circle of radius R (a simplified skull). The object is illuminated with radiation of known intensity I_0. As the beam of radiation traverses the object it is to some extent absorbed and the emergent beam is detected on the far side of the object (see Figure 2.12).

We assume that the radiation absorption coefficient of the object, $f(x,y)$, varies from point to point in the object. The physical composition of the object gives rise to this coefficient and hence a map of the coefficient would presumably give a useful profile of the object itself. The absorption coefficient governs the rate at which the radiation energy is absorbed and satisfies *Bouger's law*

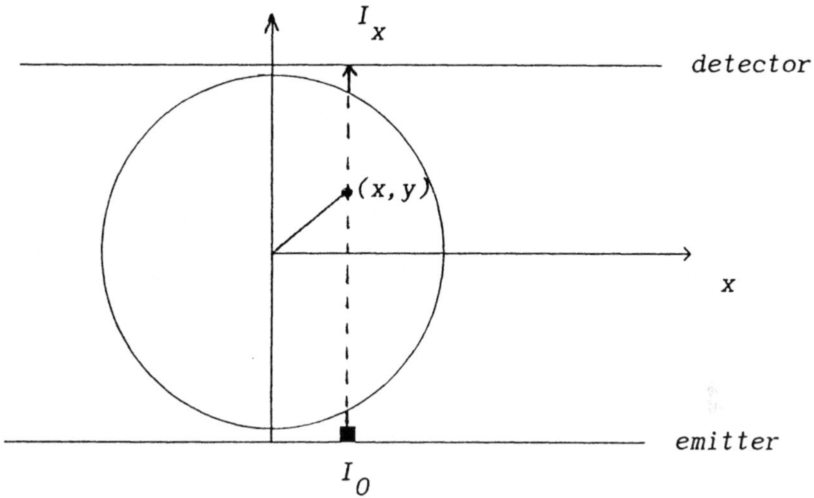

Figure 2.12

$$\frac{dI}{dy} = -fI$$

where I is the intensity of the radiation. Indeed, we may take this equation as the definition of the absorption coefficient. We then have

$$I_x = I_0 \exp\left(- \int_{-y(x)}^{y(x)} f(x, y) dy \right)$$

where $y = \sqrt{R^2 - x^2}$. Setting $p(x) = ln(I_0/I_x)$, we then have

$$p(x) = \int_{-y(x)}^{y(x)} f(x, y) dy. \tag{2.13}$$

Suppose now that f is circularly symmetric, that is, $f(x, y) = f(r)$, where $r = \sqrt{x^2 + y^2}$. Then the integral equation (2.13) becomes

$$p(x) = \int_x^R \frac{2r}{\sqrt{r^2 - x^2}} f(r) dr, \tag{2.14}$$

that is, the absorption coefficient is the solution of an integral equation of the first kind, in which the known term is the log-ratio of the radiation intensities.

Exercise 2.18: Show that the change of variables $z = R^2 - r^2$, $\tau = R^2 - x^2$, $\varphi(z) = f(\sqrt{R^2 - z})$, $P(\tau) = p(\sqrt{R^2 - \tau^2})$ converts equation (2.14) in to the Abel integral equation

$$P(\tau) = \int_0^\tau \frac{\varphi(z)}{\sqrt{\tau - z}} dz. \qquad \square$$

Stellar Stereography. Ancient astronomers were aware of diffuse luminous spots in the night sky which remained fixed relative to the stars. After the invention of the telescope these nebulae were resolved and recognized to be star clusters. Consider such a cluster which is spherical in shape, that is, a *globular cluster*. When a globular cluster is observed through a telescope, what is seen is a projection of the three-dimensional structure onto the two-dimensional focal plane of the instrument. The problem of stellar stereography consists of attempting to reconstruct the three-dimensional spherical structure of the cluster from the observed projection.

Suppose that the density of stars, per unit volume, in the spherical cluster is $x(r)$, where r is the distance from the center. Let $y(s)$ be the density of observed stars, per unit area, at a distance s from the center of the projection. If R is the radius of the globular cluster, we find on referring to Figure 2.13, that

$$\begin{aligned} y(s) &= 2\int_0^{\sqrt{R^2-s^2}} x(r)dz \\ &= \int_s^R \frac{2r}{\sqrt{r^2 - s^2}} x(r)dr \end{aligned}$$

which is the same integral equation derived in the previous example.

Exercise 2.19: Suppose that $x(r)$ represents the relative number of stars in the shell of the globular cluster contained between the radii r and $r + dr$, and that $y(s)$ represents the relative number of stars in the annulus between s and $s + ds$ in the projection. Show that

$$y(s) = \int_s^R \frac{s}{r\sqrt{r^2 - s^2}} x(r)dr. \qquad \square$$

Image Reconstruction. Most teachers are familiar, in an admittedly informal way, with the two-dimensional reconstruction problem associated with the common overhead projector. The projector consists of a light table on which a transparency is placed. Light is passed vertically through the transparency, collected by a lens arrangement and projected on an overhead screen. The light table, lens and screen may, together, be considered an *instrument* for the reconstruction of an *object*, the writing on the transparency.

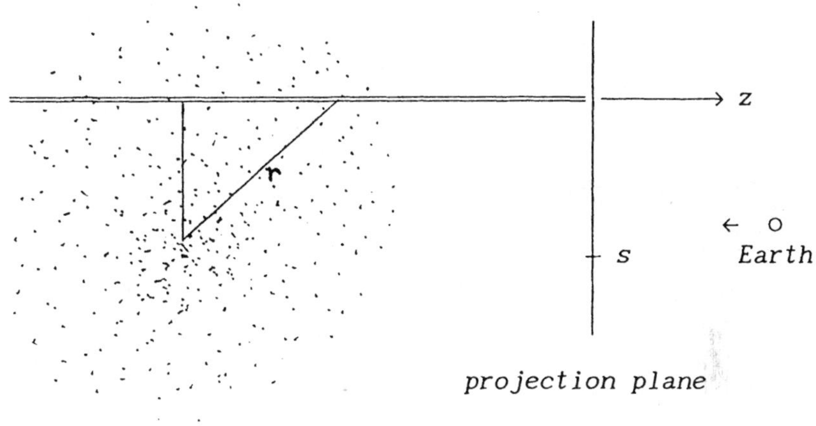

Figure 2.13

We now consider a simple one-dimensional model of such an image reconstruction process. The incident light will be a monochromatic plane wave frequency ω, which is customarily represented in complex form: $e^{i\omega t}$. The illuminated object (corresponding to writing on a transparency) is a function $f(x)$, representing a density. The situation is illustrated in Figure 2.14.

Light rays striking the object will be scattered at various angles. Consider for a moment those rays which are scattered at a fixed angle β to the horizontal. A ray scattered at point x has an optical path shortened by an amount $\Delta s = x \sin \beta$, which corresponds to a time shift of $\Delta t = \Delta s/c$, where c is the speed of light. The effect of a scattered ray at x is the shifted wave

$$f(x)e^{i\omega(t-\Delta t)} = f(x)e^{i\omega(t-\Delta s/c)} = f(x)e^{-iux}e^{i\omega t}$$

where $u = \omega\Delta s/cx = (\omega \sin \beta)/c$. Superimposing the effect of $f(x)$ over all x, we find that the scattered wave in the direction $\beta = \sin^{-1}(cu/\omega)$ is

$$\hat{f}(u)e^{i\omega t}$$

where

$$\hat{f}(u) = \int_{-\infty}^{\infty} f(x)e^{-iux}\,dx$$

is the Fourier transform of $f : \hat{f} = \mathcal{F}f$. (The notation and definition of the Fourier transform differs slightly among authors. We adopt the definition of [Pa]).

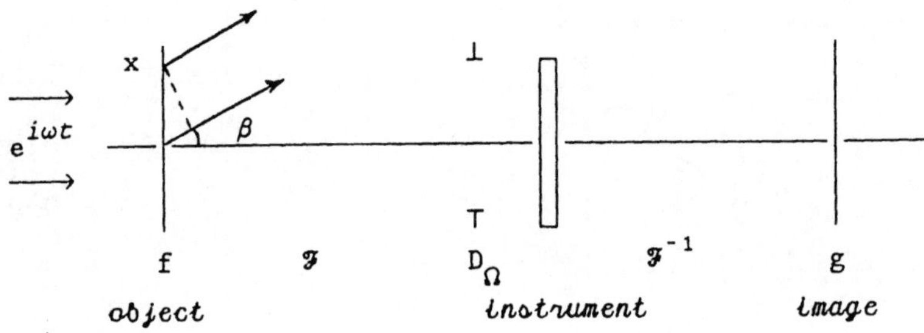

Figure 2.14

A perfect image reconstruction instrument would therefore be a realization of the inverse Fourier transform $\mathcal{F}^{-1}$. However, any real instrument is capable of collecting scattered rays for only a limited range of angles β, which corresponds to a limited range for u, say $-\Omega \le u \le \Omega$. The image g then formed by such an instrument is then modeled by

$$g = \mathcal{F}^{-1} D_\Omega \mathcal{F} f \tag{2.15}$$

where D_Ω is the low-pass filter operator

$$D_\Omega h(u) = (1_{[-\Omega,\Omega]} h)(u) = \begin{cases} h(u), & |u| \le \Omega \\ 0, & |u| > \Omega. \end{cases}$$

Now, the inverse Fourier transform of the filter function

$$1_{[-\Omega,\Omega]}(u) = \begin{cases} 1, & |u| \le \Omega \\ 0, & |u| > \Omega \end{cases}$$

is [Pa]

$$(\mathcal{F}^{-1} 1_{[-\Omega,\Omega]}(t)) = \frac{\sin \Omega t}{\pi t} = s_\Omega(t).$$

Therefore, the convolution theorem for Fourier transforms,

$$g = s_\Omega * f.$$

If the object f is known to have support contained within $[-a,a]$, i.e., if $f(x)=0$ for $|x|>a$, we then have

$$g(y)=\int_{-a}^{a}\frac{\sin\Omega(y-x)}{\pi(y-x)}f(x)dx. \tag{2.16}$$

We note that (2.16) models much more general situations than that modeled in our optical example. Suppose f is *any* signal with support contained in $[-a,a]$, which is analysed in the frequency domain by consideration of its Fourier transform $\mathcal{F}f$. If this transform is available for only a limited range of frequencies, say $[-\Omega,\Omega]$, then the available data in the time domain is

$$g=\mathcal{F}^{-1}D_{\Omega}\mathcal{F}f$$

which is again (2.16).

Exercise 2.20: Show that as an operator on $L^2(-\infty,\infty)$, $\|\mathcal{F}^{-1}D_{\Omega}\mathcal{F}\|\leq 1$. □

Our next two models are intended to show that inverse problems, phrased as integral equations of the first kind, occur (in fact, quite frequently) in the life sciences.

Immunology. We now develop a simple model in immunology relating to the reaction of antigens with antibodies in an equilibrium state. Our aim is to derive an integral equation of the first kind for the probability density of the equilibrium constant of the antigen-antibody reaction, which we take to be a random variable.

Consider first the simplest case in which an antigen AG combines with an antibody AB to form a bound antigen-antibody complex $AGAB$:

$$AG+AB \underset{k_-}{\overset{k_+}{\rightleftarrows}} AGAB.$$

The dynamics of the reaction are governed by rate constants, an association rate k_+ and a disassociation rate k_-. The rate of association is taken to be proportional to the product of the concentrations of antigen and antibody, that is, the association rate is

$$k_+[AG][AB]$$

where the brackets indicate concentrations. Similarly, the disassociation rate is

$$k_-[AGAB].$$

At equilibrium, we have

$$k_+[AG][AB]=k_-[AGAB]$$

and hence, if we define the *equilibrium constant*, x, by $x=k_+/k_-$, then

$$x[AG][AB] = [AGAB]. \tag{2.17}$$

The total number of antibodies, AB_t, consists of free antibodies AB and bound antibodies $AGAB$ and the concentrations satisfy

$$[AB_t] = [AB] + [AGAB].$$

Substituting for $[AB]$ in (2.17), we obtain

$$x[AG]([AB_t] - [AGAB]) = [AGAB]$$

and hence

$$\frac{[AGAB]}{[AB_t]} = \frac{x[AG]}{1 + x[AG]}. \tag{2.18}$$

The left hand side of this equation is the fraction of antibody molecules in the bound state. To simplify notation, we will denote the concentration of free antigen by h, i.e., $h = [AG]$. If we denote the number of antigen molecules bound per molecule of antibody by $\nu(h)$, then, assuming the antibody molecules are n-valent (i.e., that each antibody molecule has n receptor sites at which antigen molecules attach), we have

$$n[AGAB] = \nu(h)[AB_t]$$

and hence by (2.18)

$$\frac{xh}{1 + xh} = \frac{\nu(h)}{n}.$$

Finally, we suppose that the equilibrium constant x is actually a random variable with probability density $p(x)$, then, interpreting $\nu(h)$ as the average number of bound antigen molecules per antibody molecule, we have

$$\int_0^\infty \frac{xh}{1 + xh} p(x) dx = \frac{\nu(h)}{n}. \tag{2.19}$$

This Fredholm integral equation of the first kind for the probability density $p(x)$ is called the antigen binding equation. The quantity $\nu(h)$ can be determined experimentally for various concentrations h and the goal is to find the density $p(x)$. Note that, as a probability density, p must satisfy in addition (2.19) the constraints

$$0 \leq p(x) \leq 1, \qquad \int_0^\infty p(x) dx = 1.$$

Exercise 2.21: Show that the change of variables $h = e^{-s}$, $g(s) = \nu(e^{-s})/n$, $x = e^t$, $f(t) = e^t p(e^t)$ transforms equation (2.19) into the integral equation

$$\int_{-\infty}^\infty (1 + \exp(s - t))^{-1} f(t) dt = g(s). \qquad \Box$$

Permeable Membranes. We now take up a simple biologically motivated inverse problem concerned with transport across a permeable membrane. Consider a simplified one-dimensional model in which a membrane, $x = 0$, separates a compartment, $x < 0$, into which a chemical is injected, from another inaccessible compartment, $x > 0$. The concentration of the chemical, $c(x,t)$, is assumed to depend only on the single space variable x and time t. The concentration is assumed to diffuse in the same manner as heat, that is, with suitable scalings and normalizations

$$\frac{\partial^2 c}{\partial x^2} = \frac{\partial c}{\partial t}, \quad x < 0, \quad 0 < t.$$

We assume that initially the region $x < 0$ is free of the chemical, i.e., $c(x,0) = 0$ for $x < 0$, but that the concentration in the inaccessible region, $x > 0$, is some unknown function $c_+(t)$. Finally, we suppose that the transport of chemical across the membrane satisfies a law analogous to Newton's law of cooling, that is, the concentration gradient is proportional to the difference of concentrations across the membrane:

$$\frac{\partial c}{\partial x}(0,t) = k(c_+(t) - c(0,t))$$

where k is a permeability constant. The problem is to determine the function $c_+(t)$ from measurements of $c(0,t)$ on the accessible side of the membrane. We will show that $c_+(t)$ satisfies a certain Volterra integral equation of the first kind.

Let $C(x,p)$ be the Laplace transform of $c(x,t)$ with respect to variable t. From the diffusion equation and the initial condition we obtain

$$C'' = pC \tag{2.20}$$

where the primes indicate differentiation with respect to x. The gradient condition at the membrane gives

$$C'(0) = k(C_+ - C(0)). \tag{2.21}$$

We assume that the concentration is bounded for all $x < 0$ and we find from (2.20) that C has the form

$$C = A(p)e^{\sqrt{p}x}$$

for $p > 0$. The condition (21) then gives

$$A(p)\sqrt{p} = k(C_+(p) - C(0,p))$$

and hence

$$C(x,p) = k(C_+(p) - C(0,p))\frac{e^{\sqrt{p}x}}{\sqrt{p}}.$$

Applying the Laplace transform and using the convolution theorem we obtain [AS]:

$$c(x,t) = \frac{k}{\sqrt{\pi}}\int_0^t (c_+(\tau) - c(0,\tau))\frac{\exp(-x^2/4(t-\tau))}{\sqrt{t-\tau}}d\tau.$$

Letting $x \to 0$, we find that the concentration on the inaccessible side of the membrane, $c_+(t)$, satisfies the Volterra equation

$$c(0,t) = \frac{k}{\sqrt{\pi}} \int_0^t \frac{c_+(\tau) - c(0,\tau)}{\sqrt{t-\tau}} d\tau.$$

We round out our collection of models by taking up a few simplified models of inverse problems connected with radiation in various contexts. Each of the models gives rise to a Fredholm integral equation of the first kind.

Fourier Spectroscopy. When sufficiently heated, a material incandesces and radiates energy in the form of heat and light. The power dissipated as incandescent radiation is distributed over a range of wavelengths and a spectrometer may be used to analyse the power spectrum, that is, the distribution of power over various wavelengths. Figure 2.15 is a simplified picture of a spectrometer based on an interferometer

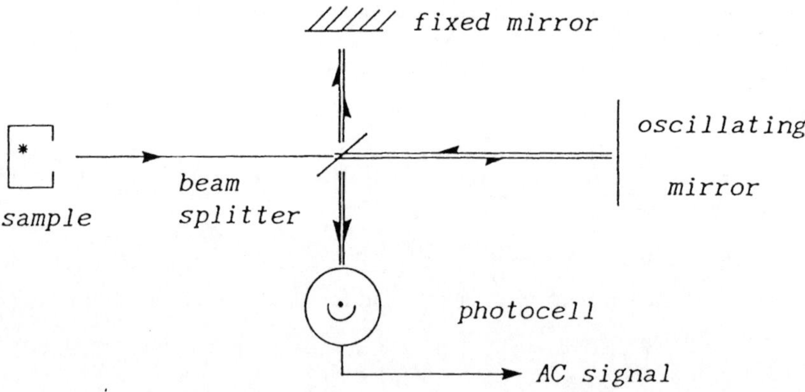

Figure 2.15

In the interferometer, a beam leaves the sample and strikes a beam-splitter (a pane of half-silvered glass) which sends part of the beam vertically and allows part to pass through. The vertical beam is reflected off a fixed mirror and the horizontal beam is reflected off an oscillating mirror (whose equilibrium position is such that its distance from the beam splitter is the same as the distance from the fixed vertical mirror to the beam splitter). The beams are recombined at the bottom of the splitter and the recombined beam enters a photocell where it is converted into an electrical signal.

Suppose that the amplitude of the original beam is A_0. If the oscillating mirror is motionless and in its equilibrium position, then the split beams recombine in phase at the beam splitter and the amplitude of the received signal at the photocell is also A_0. If the oscillating mirror is indeed in motion, then the recombined beams will generally be out of phase due to a difference δ in the length of the path followed by the two parts of the split beam. Suppose that the original beam is monochromatic with wavelength λ. The amplitude of the received signal will be a maximum when the path difference δ is an integral multiple of λ and the received signal may be represented by

$$\begin{aligned} A &= A_0 \cos(2\pi\delta/\lambda) \\ &= A_0 \cos(\sigma\delta) \end{aligned}$$

where $\sigma = 2\pi/\lambda$ is the wave number. The intensities of the signals ($I = A^2$) are then related as

$$I = I_0 \cos^2(\sigma\delta) = \frac{I_0}{2}[1 + \cos(2\sigma\delta)].$$

Now the path difference δ is a function of time when the moveable mirror oscillates (say by attaching it to an acoustical speaker and putting a pure tone through the speaker) and hence the intensity of the received signal is a function $I(t)$ of time given by

$$I(t) = \frac{I_0}{2}[1 + \cos(2\sigma\delta(t))].$$

The constant $I_0/2$ is a baseline which can be subtracted out by setting $g(t) = 2I(t) - I_0$ and hence

$$g(t) = I_0 \cos(2\sigma\delta(t)). \tag{2.22}$$

Finally, consider the original beam to be composed of signals of (infinitely) many wavelengths. The density per unit wave number, $f(\sigma)$, of the signal intensity is called the *power spectrum* and gives information on the distribution of the total power in the signal over the various wavelengths. From (2.22) we obtain

$$g(t) = \int_0^\infty f(\sigma) \cos(2\sigma\delta(t))d\sigma.$$

Radiotherapy. In this example we consider the possibility of designing a radiation treatment for tumors by implanting a metal disk, doped with a radioactive isotope, to irradiate a tumor over a long term with low dose radiation. The simplified situation we treat is this: distribute the isotope over the disk of radius R in a radially symmetric fashion in such a way that the radiation dosage on a plane parallel to the disk and at a distance $a > 0$ from the disk has a specified distribution.

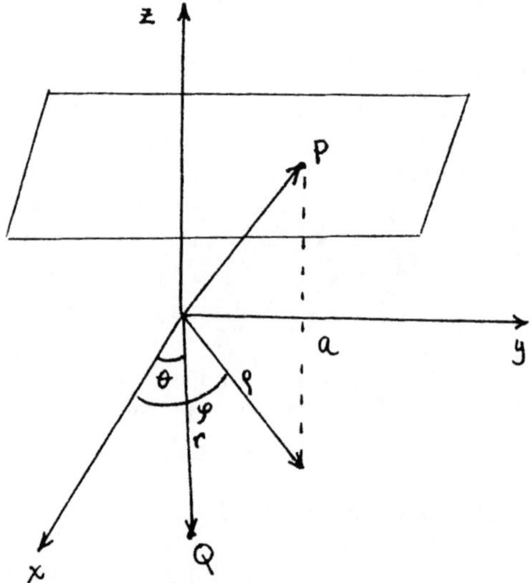

Figure 2.16

We take the disk Ω to be the disk in the xy-plane centered on the origin with radius R. The desired density of isotope on the disk will be denoted by $\sigma(q)$, $q \in \Omega$. The situation is illustrated in Figure 2.16.

The dosage at point p on the plane has the form

$$D(p) = \gamma \int_\Omega \frac{\sigma(q)}{|p-q|^2} dA$$

where γ is a constant and dA is the area element in Ω. Introducing variables as indicated in Figure 2.16, we find

$$|p-q|^2 = a^2 + r^2 + \rho^2 - 2r\rho\cos(\varphi - \theta)$$

and hence

$$D(\varphi, \rho) = \gamma \int_\Omega \int \frac{r\sigma(r)drd\theta}{r^2 + \rho^2 + a^2 - 2r\rho\cos(\varphi - \theta)}.$$

Now,

$$\int_0^{2\pi} \frac{d\theta}{r^2 + \rho^2 + a^2 - 2r\rho\cos(\varphi - \theta)} = \int_{\varphi - 2\pi}^{\varphi} \frac{d\alpha}{b - c\cos\alpha}$$

where $b = r^2 + \rho^2 + a^2$, $c = 2r\rho$.

Exercise 2.22: Show that $\int_{\varphi-2\pi}^{\varphi} \frac{d\alpha}{b - c\cos\alpha} = 2\pi/\sqrt{b^2 - c^2}$ (Hint: Use residues). □

From the exercise we have

$$\int_0^{2\pi} \frac{d\theta}{r^2 + \rho^2 + a^2 - 2r\rho\cos(\varphi - \theta)} = 2\pi \left\{(r^2 + \rho^2 + a^2)^2 - 4r^2\rho^2\right\}^{-1/2}$$

and hence this synthesis problem is modeled by the following Fredholm integral equation of the first kind:

$$D(\rho) = 2\pi\gamma \int_0^R \frac{r\sigma(r)dr}{\{(r^2 + \rho^2 + a^2)^2 - 4r^2\rho^2\}^{1/2}}. \tag{2.23}$$

Exercise 2.23: Show that the change of variables $s = r^2/R^2$, $t = \rho^2/R^2$, $d = a^2/R^2$ transforms equation (2.23) into an integral equation of the form

$$g(t) = \int_0^1 \left\{(s + t + d)^2 - 4st\right\}^{-1/2} f(s)ds. \qquad \Box$$

Black Body Radiation. A black body is an idealized physical object that absorbs all of the radiation falling upon it. When such a body is heated it emits thermal radiation from its surface at various frequencies. The distribution of thermal power, per unit area of radiating surface, over the various frequencies is called the power spectrum of the black body. The power radiated by a unit area of surface at a given frequency ν depends on the absolute temperature T of the surface and is given in appropriate units by Planck's law:

$$P(\nu) = \frac{2h\nu^3}{c^2}\frac{1}{\exp(h\nu/kT - 1)}$$

where c is the speed of light, h is Planck's constant and h is Boltzmann's constant.

Suppose that different patches of the surface of the radiating black body are at different temperatures. If $a(T)$ represents the area of the surface which is at temperature T, that is, $a(\cdot)$ is the area-temperature distribution of the radiating surface, then the total radiated power at frequency ν, $W(\nu)$, is given by

$$W(\nu) = \frac{2h\nu^3}{c^2}\int_0^\infty (\exp(h\nu/kT - 1))^{-1}a(T)dT. \tag{2.24}$$

The inverse problem of black body radiation is to find the area-temperature distribution $a(\cdot)$ that can account for an observed power spectrum $W(\cdot)$, that is, to solve the integral equation (2.24).

Exercise 2.24: Change variables in (2.24) by introducing $u = h/kT$ (the "coldness") and let $w(\nu) = c^2W(\nu)/(2h\nu^3)$. Show that (2.24) becomes

$$w(\nu) = \int_0^\infty (e^{u\nu} - 1)^{-1} a(u) du.$$

Show that w is formally the Laplace transform of

$$f(u) = \sum_{n=1}^{\infty} \frac{1}{n} a(u/n). \qquad \square$$

Atmospheric Profiling. The advent of artificial earth satellites in the late fifties afforded an unprecedented opportunity for detailed investigation of the atmosphere. Of particular interest was the temperature profile of the atmosphere, that is, the variation of temperature with altitude. Currently, gaseous profiling, particularly the determination of the ozone distribution in the atmosphere, is a matter of some urgency.

A remote sensing method of estimating the temperature profile of the atmosphere is based on a collection of microwave signals transmitted from a satellite. The physical basis for the method relies on the fact that the microwave radiation is absorbed by molecules. The excited molecules then re-emit radiation, at a rate dependent upon the temperature, according to a known physical law (Planck's law).

The idea of the method is to infer the temperature profile from measurements of the emergent intensity of radiation at the base of the atmosphere. Imagine a beam of microwave radiation at frequency ν transmitted from a satellite at an angle θ to the vertical (see Figure 2.17).

Consider what transpires as the beam traverses a layer of the atmosphere of thickness Δz at a depth z. If we assume a constant absorption coefficient k, then by Bouger's law (see the example on simplified tomography, above) the decrease in the beam intensity is given approximately by

$$kI\Delta d = \frac{k}{\mu} I \Delta z$$

where $\mu = \cos\theta$. The rate of emission of radiant energy, if the temperature in the layer is $T = T(z)$, is given by Planck's law

$$b(\nu, T) = a\nu^3/(e^{b\nu/T} - 1)$$

where a and b are constants. Combining the absorption and emission effects we have

$$\frac{\mu}{k}\frac{dI}{dz} = -I + B \tag{2.25}$$

where $B = \frac{\mu}{k} b$. Let $kz = \tau$ and write $I = I(\nu, \tau)$ to emphasize the dependence of I on frequency. From (2.25) we then obtain

$$I(\nu, \bar{\tau}) = I_0(\nu, 0)e^{-\bar{\tau}/\mu} + \frac{1}{\mu}\int_0^{\bar{\tau}} B(\nu, T(\tau)) e^{(\tau - \bar{\tau})/\mu} d\tau.$$

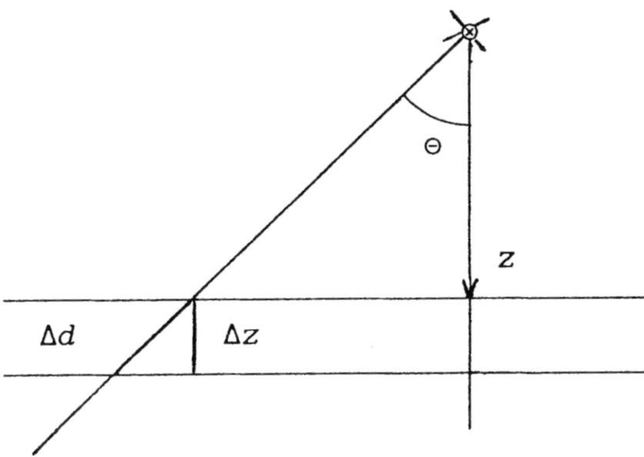

Figure 2.17

Assuming that the optical depth $\bar{\tau}$ is large enough, the first term on the right hand side of this equation is below the error threshold in measurement of the radiance $I(\nu, \bar{\tau})$ and hence is commonly disregarded yielding

$$I(\nu) = \frac{1}{\mu} \int_0^{\bar{T}} B(\nu, T(\tau)) e^{(\tau - \bar{\tau})/\mu} d\tau.$$

This is a nonlinear Fredholm equation of the first kind for the temperature profile $T(\tau)$. A great deal of work, involving both linearizations and nonlinear iterative methods for this equation, has been (and is being) carried out by the atmospheric sciences community.

2.2 Integral Equations of the First Kind

In this brief section we point out some peculiarities of integral equations of the first kind. Of course, this type of equation is the common mathematical model for all of the inverse problems introduced in the previous section and therefore the general features of such equations should be kept ever in mind in our coming discussions. For a basic introduction to integral equations of the first kind we can do no better than to recommend Wing's *Primer* [Wi]. Our aim is simply to call attention to the issues of *existence, uniqueness* and *stability*. We will have more to say on the subject in Chapter 4.

Consider a Fredholm integral equation of the first kind:

$$\int_0^1 k(s,t)x(t)dt = y(s), \qquad 0 \le s \le 1. \tag{2.26}$$

The ranges of the variables are taken for convenience as other equations may be brought into this form by appropriate transformations. We also note that Volterra integral equations of the first kind have the form (2.26) where $k(s,t) = 0$ for $t > s$. Tricomi [Tr] has said "Some mathematicians still have a kind of fear whenever they encounter a Fredholm integral equation of the first kind." While mathematicians, as a group, are not known for their fearlessness, just what is it about (2.26) that could strike terror in their hearts? Perhaps it is that mathematicians, like many people, prefer for a problem to have three things: (a) a solution, (b) not more than one solution, and (c) a solution that changes only slightly with a slight change in the problem. In mathematical parlance these conditions are known as *existence, uniqueness* and *stability*, respectively, and a problem which possesses these three attributes is called *well-posed*, a notion first clearly enunciated by Hadamard (see e.g. [Had2]) around the turn of the century. The source of the mathematical anxiety associated with (2.26) is that, on all three counts, Fredholm integral equations of the first kind are *ill-posed.*

We address the issue of existence of solutions first. Generally one would like a solution x of (2.26) of a specified type to exist for a fairly wide class of functions y. But it is clear that the function y in (2.26) depends not only on x but also inherits, throught the variable s, certain structural and qualitative properties of the kernel k. Therefore the form of k can place sever restrictions on the form of a function y for which a solution of (2.26) exists. This can become a real issue because in essentially all of the examples of the previous section y is either a measured function, and therefore has a rather general form, or, in the case of synthesis problems, y is specified a priori without much regard to the form of the kernel. To take an extreme, yet trivial, case for illustration, suppose $k(s,t) = 1$ for all $(s,t) \in [0,1] \times [0,1]$. Then an integrable solution of (2.26) exists only if the function y is a constant.

Exercise 2.25: A kernel $k(s,t)$ is called *degenerate* if it has the form

$$k(s,t) = \sum_{j=1}^{n} S_j(s)T_j(t).$$

Show that if k is degenerate, then (2.26) has no solution if y does not lie in the span of $\{S_1, ..., S_n\}$. □

Exercise 2.26: Suppose $k(s,t) = e^{st}$. Show that (2.26) does not have a bounded integrable solution if $y(s) = | s - 1/2 |$, $0 \le s \le 1$. □

What can we say about uniqueness of solutions of (2.26)? While several of the examples in the previous section have unique solutions, in general we can not expect (2.26) to have a unique solution. Indeed, in our simplest example, i.e., $k(s,t) = 1$, it is evident that for each constant function $y(s)$ there are infinitely many functions $x(t)$ satisfying (2.26). Other spectacular examples of nonuniqueness are provided by well-known orthogonality relations. For example, if $k(s,t) = \sqrt{s}\sin \pi t$, then each of the functions

$$x(t) = \sin n\pi t, \qquad n = 2, 3, \ldots$$

is a solution of the equation

$$\int_0^1 k(s,t)x(t)dt = 0.$$

Exercise 2.27: Suppose that for $0 \leq t \leq 1$, $k(s,t) = 0$ for $0 \leq s < 1/2$ and $k(s,t) = 1$ for $1/2 \leq s \leq 1$. Show that $x(t) = 0$ and $x(t) = t - 1/2$ are both solutions of $\int_0^1 k(s,t)x(t)dt = 0$, $0 \leq s \leq 1$. □

Finally, we point out that instability is a hallmark of Fredholm integral equations of the first kind. A number of instances of such instability have been pointed out in the examples of the previous section. That the instability is fundamental, and not just a consequence of some special form of the kernel, follows from the *Riemann-Lebesgue lemma.* This result states that if the kernel is square-integrable, then

$$\int_0^1 k(\cdot,t)\sin n\pi t dt \to 0 \text{ as } n \to \infty$$

where the convergence is in the sense of the mean square norm. Therefore a significant (in the mean square sense) perturbation of the form $\sin n\pi t$ to a solution $x(t)$ of (2.26) leads, for large n, to an insignificant perturbation of the effect $y(s)$. To put it another way, very small changes in the right hand side $y(s)$ can be accounted for by large changes in the solution $x(t)$.

Exercise 2.28: Let $\epsilon > 0, n$ be a positive integer and $y(s) = \epsilon \sin ns$. Show that $x(t) = \epsilon e^n \sin nt$ is a solution of

$$\int_{-\infty}^{\infty} \frac{1}{1+(s-t)^2} x(t)dt = y(s).$$

(Hint: Express $\sin nt$ as $Im e^{int}$ and use residues). □

The special form of the Volterra integral equation of the first kind does not materially simplify the difficulties associated with existence, uniqueness and stability, as the following exercises show.

Exercise 2.29: Show that the equation $\int_0^s x(t)dt = y(s)$ has an integrable solution only if y is absolutely continuous and $y(0) = 0$. □

Exercise 2.30: Show that for every real number c, $x(t) = ct^2$ is a solution of

$$\int_0^s (3s - 4t)x(t)dt = 0. \qquad \square$$

Exercise 2.31: Consider the equation $\int_0^s x(t)dt = y(s)$. Assuming that y is absolutely continuous and $y(0) = 0$, the unique solution is $x(t) = y'(t)$. In particular, $x(t) = 0$ if $y(s) = 0$. For a given $\epsilon > 0$, let $y_\epsilon(s) = \epsilon \sin(s/\epsilon^2)$. Then $| y_\epsilon(s) | \leq \epsilon$. Show that the solution $x_\epsilon(s)$ satisfies $\max_t | x_\epsilon(t) | \geq 1/\epsilon$. □

We conclude by noticing that there is a type of Volterra equation, the Volterra integral equation of the *second kind*, which in an appropriate setting always has a unique, stable solution. This equation has the form

$$x(s) = y(s) + \int_0^s k(s,t)x(t)dt. \tag{2.27}$$

It is a standard result in the basic theory of integral equations (see e.g. [Lz]) that if $k(s,t)$ is continuous for $0 \leq t \leq s \leq 1$, and $y(s)$ is continuous for $0 \leq s \leq 1$, then (2.27) has a unique solution $x(s)$ which is continuous for $0 \leq s \leq 1$. Moreover, if $\{y_n\}$ is a sequence of continuous functions converging uniformly to y and $\{x_n\}$ is the corresponding sequence of continuous solutions of (2.27), then $\{x_n\}$ converges uniformly to x. Therefore, in the space of continuous functions, the equation (2.27) is *well-posed* in the sense of Hadamard.

There is a standard technique for reducing certain Volterra integral equations of the first kind

$$\int_0^s k(s,t)x(t)dt = y(s) \tag{2.28}$$

to Volterra integral equations of the second kind. In fact, if $k(s,t)$ and $\frac{\partial k}{\partial s}(s,t)$ are continuous for $0 \leq t \leq s \leq 1$, $y'(s)$ is continuous for $0 \leq s \leq 1$ and $k(s,s) \neq 0$ for $0 \leq s \leq 1$, then we find on differentiating (2.28) with respect to s and dividing by $k(s,s)$, that

$$x(s) + \int_0^s \left(\frac{\partial k}{\partial s}(s,t)/k(s,s)\right) x(t)dt = y'(s)/k(s,s). \tag{2.29}$$

By our previous remarks this equation is well-posed in the space of continuous functions. But note that the problem of instability persists because small changes in the right hand side of the original equation (2.28) can lead to large changes in the right hand side of equation (2.29) (see Exercise 2.31). Therefore the conversion of (2.28) to an equation of the second kind merely trades the original instability in (2.28) for the instability of the differentiation process.

Exercise 2.32: Let $\varphi(s) = \int_0^s x(\tau)d\tau$. Apply integration by parts to (2.28) to obtain the Volterra integral equation of the second kind

$$\varphi(s) - \int_0^s \left(\frac{\partial k}{\partial t}(s,t)/k(s,s)\right)\varphi(t)dt = y(s)/k(s,s).$$

Does this circumvent the instability problem in (2.28)? □

2.3 Bibliographic Notes

The hanging cable model can be found in many standard texts on integral equations, for example, [Tr]. Kellogg [Ke] is still an excellent source on direct problems relating to gravitational potential. For inverse problems of gravitational potential, see [LRS], [G] and [Z]. A little known paper of some historical interest, [Jo], treats the problem of determining an unknown law of attraction from knowledge of the total attractive force of a uniform linear density. The integral equation for the gravitational edge effect is given in [D] (see also [P]) and essentially the same equation appears in another geophysical application [We].

For a discussion of the equation for the pressure guage see [Al] and [Bau]; the inverse problem for the vibrating string is treated in a number of sources, see for example, [Gl] and [PT]. The inverse heat conduction problem and lots of other inverse problems involving heat flow are taken up in [Ca], [BBC], [Mu]. Huygen's problem (Exercise 1.14) is a standard example of Abel's equation in texts on integral equations. An interesting historical perspective on the problem can be found in [L]. For more on Abel integral equations see [GV], [Lz], [FW] and [Wi]. The weir notch problem in irrigation comes from [Br] and the integral equation for the shape of a planar gravitating body (Exercise 1.17) is presented in [Iv] (see also [CB]).

Another view of the simplified tomography problem, related to plasma diagnostics is given in [McW] (see also [CMMA] for an application to flame structures). For more on tomography see, for example, [An], [Na] and [S3]. Wicksell [W] discusses the globular cluster problem (see also [CrB]) and some related problems concerning the determination of shapes of tumors from examination of tissue sections. The image reconstruction equation has been studied extensively, see for example [AMD], [N], [Ber], [Gro3] and the references cited in these works. For background on the Fourier spectroscopy equation see [Be].

A numerical method for the antigen binding equation in immunology is developed in [Ha]. The permeable membrane problem is adapted from [LLK]. The equation for the unknown concentration at the membrane is the same as that of determining the ambient temperature history from measurements of the surface temperature of a half-space. I am indebted to Dr. Bowen Keller of Rochester, New York, for bringing the radiotherapy problem to my attention; [Ma] is an early work on the corresponding direct problem.

The inverse black body radiation problem is introduced in [Bo], where the formal Laplace transform expansion of Exercise 1.24 is given (see also, [CrB], [CL] and [SJ]). The papers in the rather obscure reference [Col] are a good introduction to the atmospheric profiling problem (see also [FZ], [HTR] and [ZN]).

There are many other linear inverse problems that may be modeled in terms of integral equations of the first kind. Among these we mention the inverse travel time problem of seismic waves [BB], [C], [GV]; problems in microscopy [Co]; inverse scattering [CK]; various inverse problems in astronomy [Brw], [CrB], [JR]; polymer science [GW], [Lee]; sediment stratigraphy [Go]; radioactivity [He]; medicine [HB]; rheology [Ho]; fracture mechanics [McI]; elasticity [SP]; chemistry [TR] and transport theory [Dr].

3 Parameter Estimation in Differential Equations: Model Identification

I am not talking about how I solved
my problems, but how I posed them.
U. Eco, Postscript to
the Name of the Rose

Here is a problem that is fairly typical of those studied in the early weeks of an elementary differential equations course:

> Groundwater, containing pollutants in a concentration of 5%, seeps at a rate of 2 gallons per hour into a 1,000 gallon cistern and the well-mixed water is drawn off at the same rate. If the initial concentration of pollutants in the cistern is 1%, what is the concentration of pollutants after five days?

Good students will dutifully set up the differential equation which models the process and, perhaps with a yawn, solve it and find the required concentration of pollutants at future times. This is a more or less standard problem in which the model is completely specified and the future effect of initial conditions is calculated - a classic *direct* problem. But how are the concentration of the polluted water seeping into the cistern and the rate of seepage determined? In the real world, the seepage rate and the concentration of pollution in the groundwater would probably be obtained by measurements of the state of the cistern itself. These parameters *specify the model* and in real applications it is often the model itself which is uncertain.

If you want to get a student's attention, try posing the corresponding *inverse problem*:

> Groundwater with an unknown concentration of pollutants seeps at an unknown rate into a cistern containing 1,000 gallons and the mixture leaks out at the same rate. Measurements show that the initial concentration of pollutants in the cistern is 1%. After one day the concentration of pollutants is 1.12% and after two days it is 1.23%. What is the concentration of pollutants in the groundwater and at what rate is the groundwater seeping into the cistern?

In the inverse problem the challenge is to identify parameters which specify the model, given an initial cause and certain effects. Notice that we have specified just enough information to give a unique solution to the identification problem. However, if only one later concentration (instead of two) had been specified, the inverse problem would have infinitely many solutions (the problem is *underdetermined*), while if more than two (inconsistent) later concentrations are specified (the problem is *overdetermined*), then there may be no solution at all.

In the inverse problem above, the parameters to be identified are a mere pair of numbers. However, it is quite often the case that the parameters are *distributed*, that is, they are not simple numbers, but rather functions of the independent and/or dependent variables. In some of the models below we will find that even in the case where the distributed parameters exist and are unique, they may be numerically unstable and hence small measurement errors can make the identification of the parameters a very difficult task.

The general type of situation that we will consider in this chapter is represented schematically in the following figure. The input to the model may be many things,

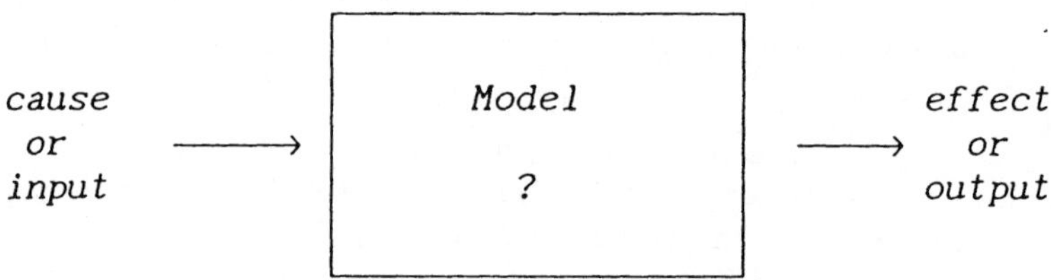

Figure 3.1

for example, initial or boundary conditions, external forcing, or even a geometrical configuration. The input is assumed to be accurately given (or generated) and the output is then measured. On the basis of these data we wish to identify certain parameters which specify the model. Some issues arise immediately. Is there a model (in some class) that can account for the given data (existence)? Is there more than one such model (uniqueness)? Is the model overly sensitive to measurement errors in the data (stability)? How does one obtain stable, reliable values for the parameters which specify the model (approximation)?

In this chapter we will not deal with these issues in detail, nor will we propose general approaches to these problems. Our aim is rather to present some concrete inverse problems of parameter identification in differential equations in the simplest terms possible. In fact, we feel that the approach we take is elementary enough to present such problems, and discuss the issues which arise, at appropriate places in the undergraduate curriculum. We consider now some elementary models involving identification of parameters in differential equations.

3.1 An Exponential Growth Model

Perhaps the most familiar mathematical model involving a differential equation is the growth law

$$\frac{du}{dt} = ru. \tag{3.1}$$

Here u might represent the population of a colony of bacteria, where r is a growth constant, or u might represent the value of an investment which is being compounded continuously at a rate r. If r is a constant, then the direct problem has the solution

$$u(t) = u_0 e^{rt}, \quad \text{where} \quad u_0 = u(0)$$

and the parameter r may be determined by measuring one later value of u, say $u(1)$:

$$r = \ln(u(1)/u_0).$$

But what if r is a function of t (e.g., variable fecundity rate; variable interest rate)? The inverse problem of determining $r(t)$ can be solved directly from the differential equation, namely

$$r(t) = \frac{du}{dt}/u = \frac{d}{dt}(\ln u). \tag{3.2}$$

So, to determine the distributed parameter $r(t)$ we simply observe the quantity u (assumed positive) and apply (3.2). Simple - or is it? Remember that the quantity $u(t)$ is measured and hence is subject to measurement errors. This can have unpleasant consequences for, as we have seen in Exercise 3.31, the process of differentiation can be quite unstable. Consider, for example, what would happen if $u(t)$ were perturbed to $u_\epsilon(t)$. Then $\ln u(t)$ would be perturbed to $\ln u_\epsilon(t)$ and if, for example,

$$\ln u_\epsilon(t) = \ln u(t) + \epsilon \sin(t/\epsilon^2).$$

where ϵ is a small positive number, then the perturbation in $\ln u(t)$ is quite small, but the perturbation in $\frac{d}{dt} \ln u(t)$, namely

$$\frac{1}{\epsilon} \cos(t/\epsilon^2)$$

is very large. Such instability is typical of inverse problems involving identification of distributed parameters.

Let's have a look at the practical implications of this instability. How do we "observe" u? A simple way would be to take a sequence of measurements $u_0, u_1, u_2, \ldots$, which are equally spaced, say with spacing $h > 0$. The measurement u_k then represents an approximation to the true value $u(kh)$, $k = 0, 1, 2, \ldots$. Setting $f(t) = \ln u(t)$, we can find approximations to the rate coefficient $r(t)$ by making an approximation to the derivative in (3.2), for example,

$$\frac{d}{dt} f(kh) \approx \frac{f((k+1)h) - f(kh)}{h}.$$

Writing r_k for an approximation to $r(kh)$, we are led to the following approximation scheme for values of the distributed coefficient:

$$r_k = \frac{\ln u_{k+1} - \ln u_k}{h}. \tag{3.3}$$

Assuming that $u(0) > 0$, the approximations are consistent in the sense that for any fixed $t > 0$, if $h = t/k$ and if the measurements are exact, that is, $u_k = u(kh)$, then

$$\lim_{k \to \infty} r_k = \frac{d}{dt} \ln u(t) = r(t).$$

But what is the effect of inevitable measurement errors? Suppose that the actual measurements satisfy

$$| f_k - f(kh) | \leq \epsilon \tag{3.4}$$

where $f_k = \ln u_k$ and $f(t) = \ln u(t)$. We then have for $t = kh$.

$$\begin{aligned} | r(t) - r_k | &= | f'(t) - (f_{k+1} - f_k)/h | \\ &\leq | f'(t) - (f(t+h) - f(t))/h | \\ &\quad + | f((k+1)h) - f_{k+1} - f(kh) + f_k | /h. \end{aligned}$$

If we assume that f has a continuous second derivative, then by Taylor's theorem the first term in the upper bound for the error above is bounded (for t in a closed bounded interval) by a constant times h. We then say that the term is "order of h" and write it as $O(h)$. From (3.4), the second term is bounded by $2\epsilon/h$. Therefore

$$| r(t) - r_k | \leq O(h) + 2\epsilon/h = O(1/k) + kO(\epsilon) \tag{3.5}$$

where $t = kh$ is fixed. This bound illustrates the classic stability dilemma: for a fixed error level $\epsilon > 0$, the first term (truncation error) goes to zero, but the second term (stability error) blows up, as $k \to \infty$. The parameter k (or equivalently h) controlling the approximation process may not be chosen arbitrarily, but must be carefully tailored to the model and the errors in the observations. This is a theme that will recur frequently in our subsequent discussion of unstable inverse problems.

Although it is not clear what a "best" choice of the discretization parameter h in this scheme would be, it is plain to see that a balancing of competing terms in (3.5) would suggest a form $h = c\sqrt{\epsilon}$. With such a choice we find that

$$| r(t) - r_k | = O(\sqrt{\epsilon}),$$

that is, the resolution we can expect in the computed values of the rate coefficient is on the order of the square root of the errors with this choice of the discretization parameter h.

Exercise 3.1: Show that if f has a continuous third derivative and the approximation

$$f'(t) \approx \frac{f(t+h) - f(t-h)}{2h}$$

is used, then a choice of discretization parameter of the form $h = C\epsilon^{1/3}$ leads to an approximation for the rate coefficient of order $O(\epsilon^{2/3})$ when the function values $f(t)$ are subject to errors of order ϵ. □

Exercise 3.2: Write a computer program to carry out the approximation scheme (3.3). As a test case, consider $u(t) = \exp(t^2)$ on $[0,1]$, which satisfies (3.1) with $r(t) = 2t$ and $u(0) = 1$. Run your program for various choices of the discretization h, using $u_k = \exp((kh)^2)$ as data and compare the computed values r_k with the true values $r(kh) = 2kh$. Then take a small positive number ϵ and generate random perturbations in $[-\epsilon, \epsilon]$, adding them to the u_k. Calculate the approximations r_k using these noisy measurements and compare with the true values for a range of mesh sizes h. Observe the effect of the instability in the process on the computed approximations. □

Exercise 3.3: Another algorithm for $r(t)$ could be based on integrating equation (3.1) from t_k to t_{k+1} and approximating the resulting right hand side by the trapezoidal rule, that is,

$$u_{k+1} - u_k = \frac{h}{2}\left[r_k u_k + r_{k+1} u_{k+1}\right]$$

or equivalently,

$$r_{k+1} = \left[2(u_{k+1} - u_k)/h - r_k u_k\right]/u_{k+1}.$$

Of course, this method requires an initial approximation r_0, which could be obtained from (3.3). Write a computer program based on this algorithm. Use the program on the same test case as in the previous exercise and investigate the effect of noisy measurements on the calculations. □

3.2 A Problem in Hydraulics

In the previous example the unknown coefficient, $r(t)$, depended only on the independent variable. We now consider a simple example of a one-dimensional coefficient identification problem in which the distributed parameter is a function of the dependent variable. Imagine a vessel with irregular cross section containing water that is draining through a hole, of cross sectional area a, at its base.

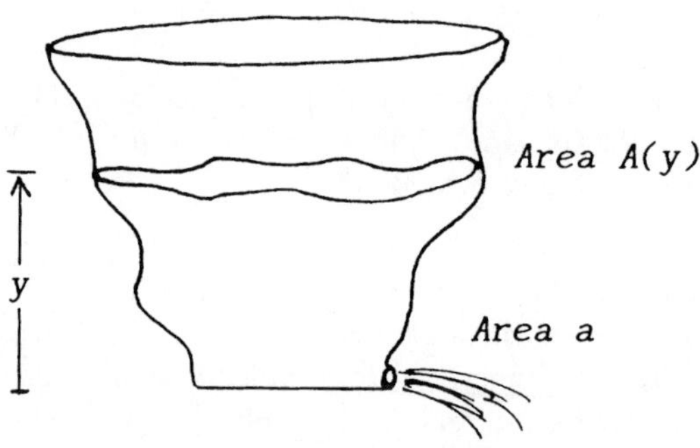

Figure 3.2

If the shape, or in fact just the cross-sectional area profile $A(y)$, of the vessel is known, then it is a relatively simple *direct* problem to predict the water level y as a function of time. The key to the analysis is an elementary energy balancing relation in the guise of *Torricelli's Law*, which gives the velocity of the effluent.

Consider an infinitesimal volume slice $V = A(y)\Delta y$ at the surface of the water which is drained away in time Δt. There is a consequent loss of potential energy of magnitude Vgy, where g is the gravitational constant (for convenience we have taken density of the water to be one). This loss of potential energy must be balanced by the kinetic energy of this volume element as it passes through the drain. If we denote the velocity of the effluent by v, then the kinetic energy is

$$\frac{1}{2}mv^2 = \frac{1}{2}Vv^2$$

and equating this with the potential energy loss we have

$$\frac{1}{2}Vv^2 = Vgy$$

or

$$v = \sqrt{2gy}.$$

This is Torricelli's law. Consider now the total volume of water, $V(y)$, contained in the vessel. If the water drains, with velocity v, through a hole with cross-sectional area a, then

$$\frac{d}{dt}V(y) = -av(y) = -a\sqrt{2gy}.$$

However,

$$\frac{d}{dt}V(y) = V'(y)\frac{dy}{dt} = A(y)\frac{dy}{dt}$$

and hence the water level y satisfies the differential equation

$$A(y)\frac{dy}{dt} = -a\sqrt{2gy}. \tag{3.6}$$

The direct problem consists of determining the water level y from this differential equation and knowledge of the cross-sectional area $A(y)$ and the initial depth $y(0)$.

Exercise 3.4: Find the water level history $y(t)$ for the case where the vessel is: (i) A right circular cylinder of radius r; (ii) An inverted cone of height h and base radius r; (iii) A hemisphere of radius r. □

Exercise 3.5: What is the shape of a vessel that gives a constant rate of decrease for the water level? Is such a vessel unique? How would one synthesize a vessel so as to produce a given monotone decreasing water level history $y(t)$? □

This exercise points to an equally interesting *inverse* problem associated with the draining vessel. Namely, suppose that the water level y can be measured, but the cross-sectional area $A(y)$ is inaccessible. The situation is illustrated in Figure 3.3 (a sealed institutional coffee urn with unknown internal geometry is an appropriate visualization). The inverse problem of determining the cross-section $A(y)$ from the water level history $y(t)$ is then the problem of determining the distributed parameter $A(y)$ in the differential equation (3.6). Under the reasonable assumption that y is monotone decreasing, $dy/dt < 0$, we can solve explicitly for $A(y)$:

$$A(y) = \frac{-a\sqrt{2gy}}{\frac{dy}{dt}} \tag{3.7}$$

(note that in practice there is no region where $\frac{dy}{dt} = 0$ and hence $A(y)$ is in principle identifiable). Although (3.7) uniquely determines $A(y)$, we see that practical determination of $A(y)$ will be at the mercy of the instabilities inherent in the differentiation of the measured water level $y(t)$. However, if $y \geq b > 0$ (why can this always be arranged?) and $dy/dt \leq c < 0$, then given inaccurate measurements y_k satisfying

$$| y_k - y(kh) | < \epsilon$$

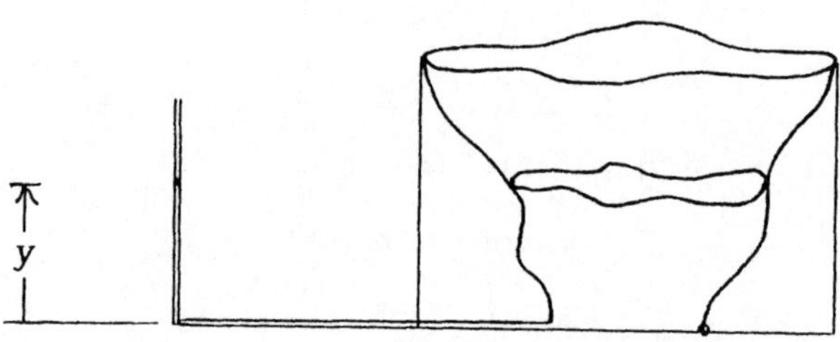

Figure 3.3

one can arrange for the approximations $A(y_k)$ given by:

$$A(y_k) = \frac{-a\sqrt{2gy_k}}{y_k - y_{k-1}}h, \; A(y(0)) = A(y_0) \tag{3.8}$$

to satisfy $A(y_k) \to A(y(t))$ as $k \to \infty$, where $t = kh > 0$ is fixed and h is appropriately related to ϵ.

Exercise 3.6: Carry out the details of the convergence discussion given immediately above (see the discussion of the previous example). □

Exercise 3.7: Write a computer program based on (3.8) to calculate values of the coefficient $A(y)$ from measured values of y. Test your program, using "clean" data on the examples of Exercise 3.4 where y and $A(y)$ are both known. Investigate the performance of the algorithm when the data y_k are perturbed with random errors. □

Exercise 3.8: Show that integrating both sides of (3.6) from t_k to t_{k+1} and using the trapezoidal rule suggests the method

$$A(y_{k+1}) = -a\sqrt{2g}\frac{h}{\sqrt{y_{k+1}} - \sqrt{y_k}} - A(y_k).$$

Repeat the previous exercise using this method and compare results. □

3.3 Compartmental Analysis

We now take up a class of inverse problems involving the determination of constant coefficients in systems of ordinary differential equations. To set the scene, consider the problem of coefficient identification in a linear algebraic system. Given a vector $x \in \mathbf{R}^n$ and the vector $b = Ax$ in $\mathbf{R}^m$, one wishes to determine the $m \times n$ matrix A. This is of course a severely underdetermined situation and no unique matrix A

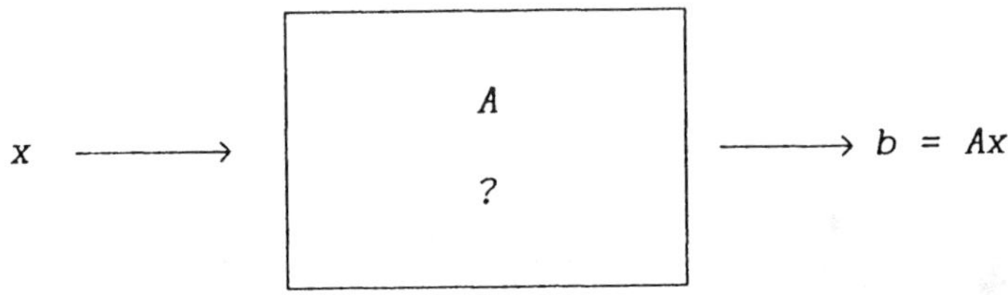

Figure 3.4

can be identified satisfying these conditions. However, if we are allowed the luxury of specifying more inputs and observing the corresponding outputs, then A can be easily identified. For example, if we choose successively for x the n standard basis vectors, then the corresponding vectors b are the columns of A.

Exercise 3.9: Show that n linearly independent vectors x, and the corresponding vectors b, serve to identify the $m \times n$ matrix A. □

In the models to follow in this section, the identification problem is complicated by dynamic factors, that is, the appropriate quantities for identifying the system are *derivatives* of measured data. Consequently, the stability problems associated with differentiation of measured data will in one way or another make themselves known.

Compartmental modeling is a powerful technique for analyzing biological and biochemical processes and structures. The process is represented by a system of first order ordinary differential equations with constant coefficients in which the coefficients represent structural parameters of interest and the solution is a vector of measurable dynamic quantities. Among the applications of compartmental analysis are studies of metabolism, chemical kinetics, renal function, tracer diagnostics and many other biological and physical processes.

In compartmental analysis the system is investigated by lumping it into a finite number of distinct compartments which exchange material between one another (it is helpful to think of a compartment as a well-defined physical vessel, however, in general a "compartment" can consist of a conceptually distinct quantity of material without a well-defined geometry). The exchange of material between compartments may be driven by diffusion, thermal gradients, chemical or biological reactions or other mechanisms. Material within any given compartment is considered to be well-mixed and homogeneous.

As a first example, consider a simple two compartment model for ingestion and metabolism of a drug. As compartments we choose the gastrointestinal tract (compartment 1) and the bloodstream (compartment 2). We assume that the drug is ingested at a rate $i_1(t)$ and is exchanged between compartments (or eliminated in the environment) at rates which are proportional to the concentration of drug in a given compartment. The constants of proportionality are called *transfer rates* and it is these rates that are our primary concern. For example, if the two compartment model is represented as in Figure 3.5, then the governing system of differential equations is

$$\begin{aligned} \frac{dx_1}{dt} &= i_1(t) - k_{21}x_1(t) \\ \frac{dx_2}{dt} &= k_{21}x_1(t) - k_{02}x_2(t) \end{aligned}$$

where $x_i(t)$ represents a concentration and k_{ji} is a transfer rate from compartment i to compartment j. Note that in this example we are assuming a one-way transfer from compartment 1 to compartment 2. In matrix notation the system takes the form

$$\dot{x} = Ax + u$$

where

$$x = \begin{bmatrix} x_1 \\ x_2 \end{bmatrix}, \quad A = \begin{bmatrix} -k_{21} & 0 \\ k_{21} & k_{02} \end{bmatrix} \quad \text{and} \quad u = \begin{bmatrix} i_1 \\ 0 \end{bmatrix},$$

and the dot indicates differentiation with respect to time. The inverse problem of compartmental analysis consists of determining the matrix of transfer rates A from knowledge of u and measurements of x.

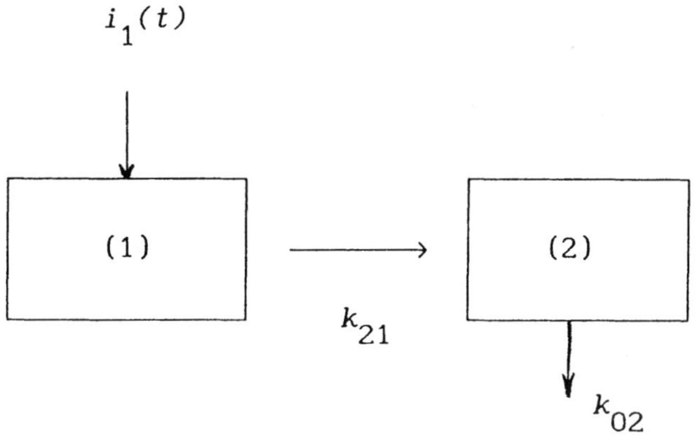

Figure 3.5

Exercise 3.10: Find k_{21} and k_{02} in the above model if $i_1(t) = 3e^{2t}$, $x_1(t) = x_2(t) = e^{2t}$. □

Exercise 3.11: Show that no coefficients k_{02}, k_{21} exist which would result in the response $x(t) = (e^{-t} - 1)[1\ ,\ 1]^T$. □

Exercise 3.12: In this exercise you are asked to develop a system of differential equations modeling transfer between two compartments separated by a permeable membrane, as in Figure 3.6. We assume that *Fick's law* governs transfer across the membrane, that is, the rate of transfer of material across the membrane is proportional to the product of the area of the membrane with the concentration difference of the compartments (see also the membrane model in Chapter I). We take the cross sectional area of the membrane to be 1 and denote the permeability coefficient for flow from compartment i to compartment j by k_{ji}. Then Fick's law for the amount of material y_1 in compartment 1 is

$$\frac{dy_1}{dt} = -k_{21}c_1 + k_{12}c_2$$

where $c_i(t)$ is the concentration of material in compartment i. Suppose that the (constant) volumes of compartments 1 and 2 are V_1 and V_2, respectively. Show that the system in Figure 3.6 is modeled by

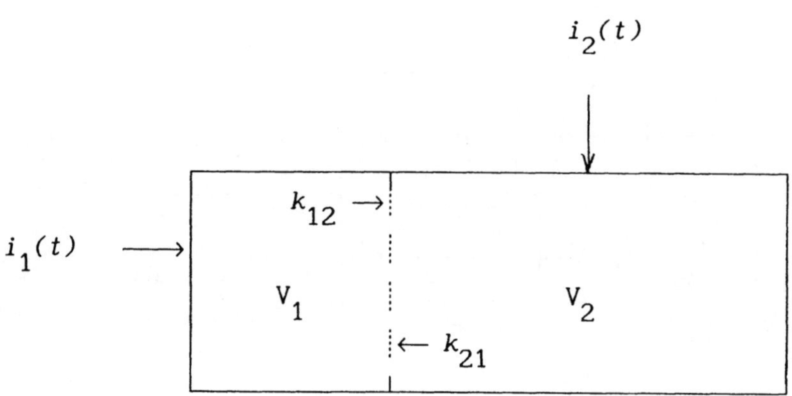

Figure 3.6

$$\dot{y} = A\dot{y} + u$$

where

$$A = \begin{bmatrix} -k_{21} & k_{12} \\ k_{21} & -k_{02} - k_{12} \end{bmatrix} \begin{bmatrix} V_1 & 0 \\ 0 & V_2 \end{bmatrix}$$

and $u = [i_1 \ , \ i_2]^T$. □

In the examples above, and in much more complicated compartmental models, the dynamics of the system are described by a system of ordinary differential equations of the form

$$\dot{x}(t) = Ax(t) + u(t)$$

where A is an $n \times n$ matrix, $u(t) = [u_1(t), ..., u_n(t)]^T$ is an *input* (or *control*) function and $x(t) = [x_1(t), ..., x_n(t)]^T$ is the solution or *response* function. In tracer diagnostics the input $u(t)$ typically consists of a single or a few, controlled injections, that is,

$$u(t) = Bw(t)$$

where $w(t) = [w_1(t), ..., w_k(t)]^T$ represents a small number, k, of inputs and B is an $n \times k$ input distribution matrix that represents how the inputs are distributed among the compartments of the system. Similarly, the response is sampled via an output

$y(t)$ of the form $y(t) = Cx(t)$, where C is a $p \times n$ output sampling matrix which determines which compartments are tapped for measurements. We will assume zero initial conditions. The dynamics of the compartmental system are then modeled by

$$\begin{aligned} \dot{x}(t) &= Ax(t) + Bw(t), \quad t > 0 \\ x(0) &= 0 \end{aligned} \tag{3.9}$$

and the response is sampled by

$$y(t) = Cx(t).$$

For example, in the three compartmental model in Figure 3.7, measurements are

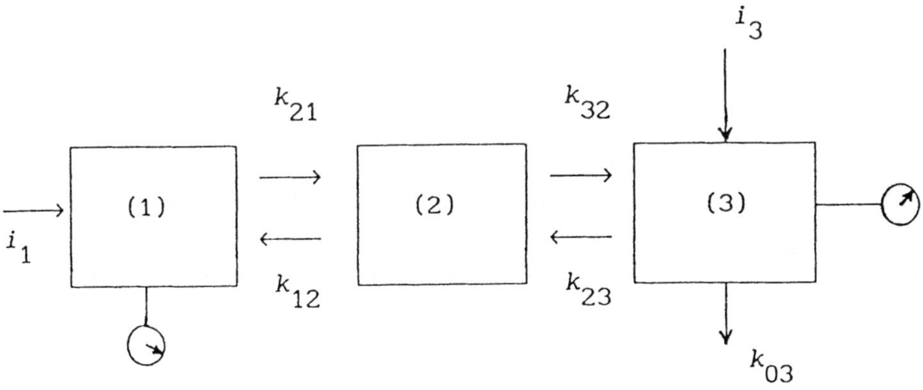

Figure 3.7

indicated by the gauge symbol and the output sampling function is

$$y(t) = Cx(t) \text{ where } C = \begin{bmatrix} 1 & 0 & 0 \\ 0 & 0 & 1 \end{bmatrix}.$$

The input function is

$$u(t) = Bw(t) \text{ where } w = [i_1 \ , \ i_2 \ , \ i_3]^T \text{ and } B = \begin{bmatrix} 1 & 0 & 0 \\ 0 & 0 & 0 \\ 0 & 0 & 1 \end{bmatrix},$$

and the dynamics governed by

$$\dot{x} = Ax + Bw$$

where

$$A = \begin{bmatrix} -k_{21} & k_{12} & 0 \\ k_{21} & -k_{12} - k_{32} & k_{23} \\ 0 & k_{32} & -k_{23} - k_{03} \end{bmatrix}.$$

An explicit solution of (3.9) can be written in terms of a convolution of the input function and a matrix exponential, in fact

$$x(t) = \int_0^t e^{A(t-s)} Bw(s)ds \tag{3.10}$$

and hence the output sampling function is given by

$$y(t) = \int_0^t Ce^{A(t-s)} Bw(s)ds$$

or, in convolution notation:

$$y = Ce^{A(\cdot)} B * w. \tag{3.11}$$

Exercise 3.13: Verify that the vector function $x(t)$ given in (3.10) solves (3.9). □

The convolution formulation (3.11) immediately suggests the use of Laplace transform analysis. Indeed, applying the Laplace transform to (3.11), we find

$$Y = \Phi W \tag{3.12}$$

where the *transfer function* Φ is given by

$$\Phi(s) = C(sI - A)^{-1} B. \tag{3.13}$$

Exercise 3.14: Verify (3.13). □

From (3.11-3.12) we see that the transfer function relates the inputs to the outputs and hence it holds the key to system *identifiability*. If the inputs and outputs are known, then C, B, W and Y are known and hence we see from (3.13) that the system is identifiable if Φ determines A.

Exercise 3.15: Show that the system described in Figure 3.8 is identifiable if $k_{02} = 0$, but not identifiable if $k_{02} \neq 0$. □

The transfer function can be used, as indicated above, to analyze the theoretical problem of identifiability of the model. But how can the model parameters, that is, the matrix A, be determined numerically? A general technique that is often used for this purpose is called *output least squares*. In this method, the outputs are measured at discrete times $t_1, t_2, \ldots, t_m$. Call these measured vectors $y^{(1)}, y^{(2)}, \ldots, y^{(m)}$. If these measurements were exact, then we would have

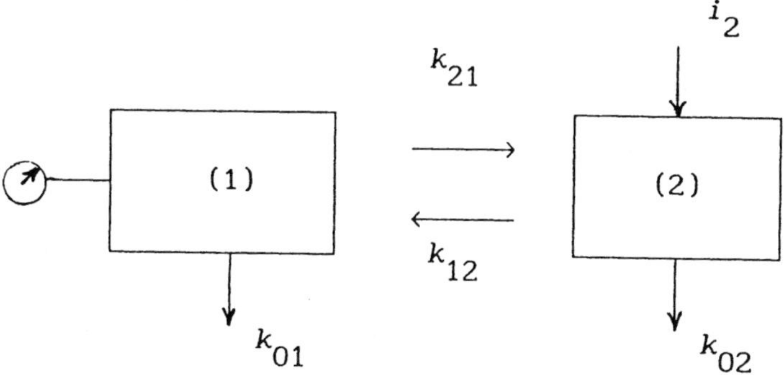

Figure 3.8

$$y^{(i)} = \int_0^{t_i} Ce^{A(t_i-s)}Bw(s)ds, \quad i = 1, 2, ..., m$$

where A is an unknown $n \times n$ matrix. However, the measurements are not exact and in the output least squares method estimates for the unknown entries $a_{ij}, i, j = 1, ..., n$ are obtained as the n^2 values which minimize the least squares functional

$$\sum_{i=1}^{m} \| y(t_i) - y^{(i)} \|^2 . \tag{3.14}$$

The expression in (3.14) is a nonlinear function of the n^2 variables a_{ij} and its minimization is typically accomplished by some Newton-like iterative scheme.

Exercise 3.16: Verify that for the compartmental model in Figure 3.9, the solution with $x(0) = 0$ is

$$\begin{aligned} x_1(t) &= (1 - e^{-k_{21}t})/k_{21} \\ x_2(t) &= (k_{02}(1 - e^{-k_{21}t}) + k_{21}(e^{-k_{02}t} - 1))/k_{02}(k_{02} - k_{21}). \end{aligned}$$

If 50 measurements $x^{(1)}, ..., x^{(50)}$ are taken for the model, explain how the minimization of (3.14) leads to a system of two nonlinear equations for the transfer coefficients k_{21} and k_{02}. □

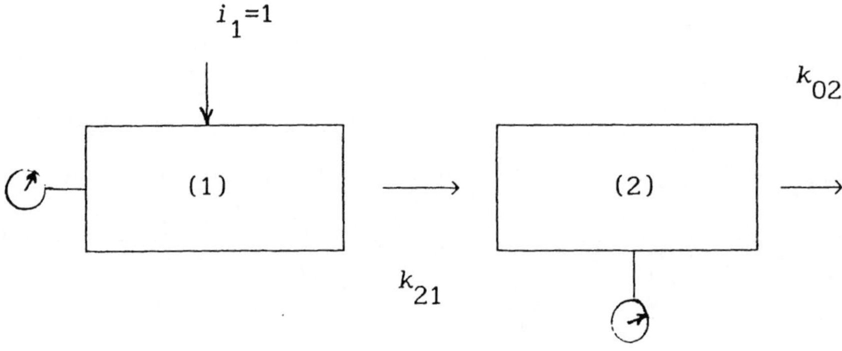

Figure 3.9

3.4 Structural Dynamics

The determination of structural parameters is big business. Even "rigid" structures vibrate when excited. In fact, the ability to vibrate is a necessary part of the design of earthquake-proof structures and knowledge of structural parameters is necessary to predict the type of excitation that can lead to destructive vibrations. Large static structures such as building and bridges, as well as smaller dynamic structures, such as helicopters or automobiles, require extensive vibration testing and analysis in the design and development stages. An important part of the process, the determination of structural parameters, involves solving an inverse problem for the coefficients in a system of differential equations.

We begin with the simplest one-dimensional situation. Consider a body of mass m suspended on a stiff spring with *Hooke's constant* k. If the body is displaced from its equilibrium position and released, it will execute an oscillatory motion which we will assume is damped by a resistive force which is proportional to the velocity. The constant of proportionality, c, will be called the *damping constant.* The situation is illustrated in Figure 3.10.

Newton's second law then gives

$$m\ddot{x} = -c\dot{x} - kx$$

or

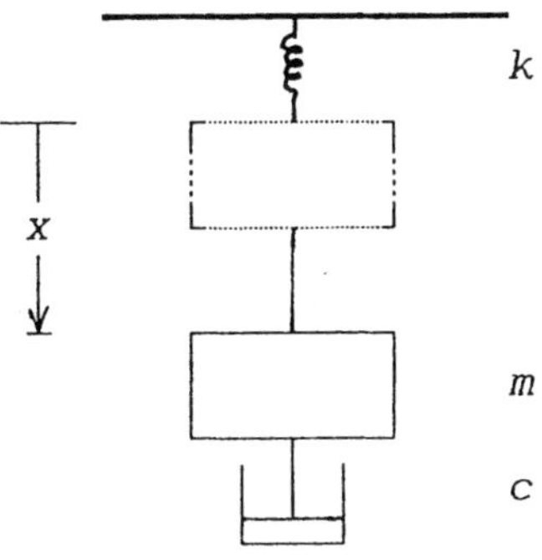

Figure 3.10

$$m\ddot{x} + c\dot{x} + kx = 0, \tag{3.15}$$

and as initial conditions we take $x(0) = x_0, \dot{x}(0) = \dot{x}_0$. The standard direct problem in an elementary differential equations course is to find the response x, given the parameters m, c, and k. We are concerned with the inverse problem of determining the mass, damping constant and stiffness constant from observations of the response x. Of course, since (3.15) is homogeneous (i.e., unforced) the best we can hope for is to determine the ratios $a = c/m$ and $b = k/m$ in the equation

$$\ddot{x} + a\dot{x} + bx = 0. \tag{3.16}$$

One way to proceed is to integrate the equation (3.16) twice using the initial conditions. This leads to

$$x(t) - x_0 - \dot{x}_0 t + a\int_0^t (x(s) - x_0)ds + b\int_0^t x(s)(t-s)ds = 0, \quad t > 0. \tag{3.17}$$

This is, of course, a severely over-determined system in that two parameters, a and b, are obliged to satisfy infinitely many conditions (one for each $t > 0$). Nevertheless, the true values of these parameters will do exactly this. However, in practice only measured values of the response will be known at a finite set of discrete times. Suppose that n measured values of x, corresponding to times $h, 2h, ..., nh$, are available. If we call the measured values $x_1, x_2, ..., x_n$ and if we approximate the integrals in (3.17) by, say, the trapezoidal rule, then for each $k = 1, 2, ..., n$ we would like the quantity

$$\begin{aligned} E_k(a,b) &= x_k - x_0 - \dot{x}_0 kh + a\left(\sum_{j=1}^{k} x_j h - x_k h/2 - x_0 kh\right) \\ &\quad + b\left(\sum_{j=1}^{k-1} x_j (k-j)h^2 + x_0 kh^2/2\right) \end{aligned}$$

to be small. The *method of least squares* for estimating the parameters a, b accomplishes this by minimizing the quantity

$$E(a,b) = \sum_{k=1}^{n} (E_k(a,b))^2.$$

Note that minimizing $E(a,b)$ is equivalent to solving a system of two linear equations for the least squares estimates of a and b.

Exercise 3.17: Solve (3.15) along with the initial conditions $x(0) = 1$, $\dot{x}(0) = -1$, with $m = 1$, $c = 2$, $k = 5$. For $t_k = kh$, where $h > 0$, generate noisy measurements $x_k = x(t_k) + \epsilon_k$, for $t_k \leq 1$, where ϵ_k is a uniformly distributed random number in $[-\epsilon, \epsilon]$ and $x(t)$ is the true solution. Write a program to estimate $a = c/m$ and $b = k/m$ by the least squares technique and compare the estimates with the true parameters $a = 2$, $b = 5$ for various choices of ϵ. Investigate the stability of the method with respect to ϵ. □

We now take up some problems with more than one degree of freedom, starting with a simple example in which no account is taken of damping. Consider two unit masses attached to a taut string and positioned as in Figure 3.11. We assume that the string is under uniform tension τ, and for simplicity we take $\tau = 1$. If the masses are drawn from their equilibrium positions and released, they will execute vertical oscillations as indicated in the figure. The picture is exaggerated in the sense that we assume the amplitudes $\mid x_1 \mid$ and $\mid x_2 \mid$ are small relative to a, b and c, and hence the angles α, β and γ are *small.* The governing nonlinear differential equations are

$$\begin{aligned} \ddot{x}_1 &= -\sin\alpha - \sin\beta \\ \ddot{x}_2 &= \sin\beta - \sin\alpha \end{aligned}$$

As we are assuming the angles are small, we have $\sin\alpha \approx \tan\alpha \approx x_1/a$ and

$$\begin{aligned} \sin\beta &\approx (x_1 - x_2)/b \\ \sin\gamma &\approx x_2/c. \end{aligned}$$

Therefore the linearized equations of motion are

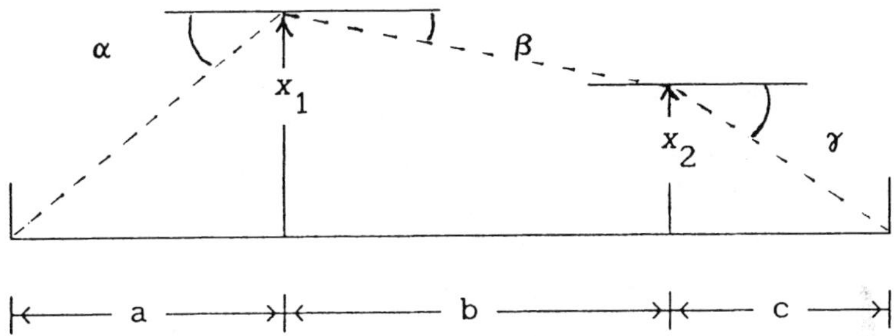

Figure 3.11

$$\begin{aligned}\ddot{x}_1 &= \left(-\frac{1}{a}-\frac{1}{b}\right)x_1+\frac{1}{b}x_2\\ \ddot{x}_2 &= \frac{1}{b}x_1+\left(-\frac{1}{c}-\frac{1}{b}\right)x_2\end{aligned}$$

or

$$\ddot{x} = Ax, \tag{3.18}$$

where

$$A=\begin{bmatrix}-\frac{1}{a}-\frac{1}{b} & \frac{1}{b}\\ \frac{1}{b} & -\frac{1}{c}-\frac{1}{b}\end{bmatrix} \text{ and } x=\begin{bmatrix}x_1\\ x_2\end{bmatrix}.$$

Imagine now that an observer, sighting along the axis of the string, can observe x_1 and x_2, but is unaware of the distances a, b and b between the masses. The inverse problem of determining the distances is then the problem of determining the matrix A from observations of x.

Exercise 3.18: Suppose that x satisfies (3.18), $x(0) = [1\ ,\ -1]^T$ and $\dot{x}(0) = [0\ ,\ 0]^T$. Show that

$$x(t) = [1\ ,\ -1]^T + A\int_0^t x(s)(t-s)ds.$$

Devise a least-squares method for estimating the matrix A from observations $x^{(1)}, ..., x^{(N)}$ of $x(h), x(2h), ..., x(Nh)$, where $h > 0$. Implement your method as a computer program and test it on randomly perturbed observations of the true solution x generated by a given matrix A. Observe the sensitivity of the matrix entries to the size of the random perturbations. □

As a final example of a multi-dimensional vibration problem, consider the pair of coupled, damped, vibrating masses in Figure 3.12. If the masses are excited by external forces $f_1(t)$ and $f_2(t)$, respectively, then the equations of motion are:

$$\begin{aligned} m_1\ddot{x}_1 &= -k_1x_1 + k_2(x_2 - x_1) - c_1\dot{x}_1 - c_2(\dot{x}_1 - \dot{x}_2) + f_1(t) \\ m_2\ddot{x}_2 &= -k_2(x_2 - x_1) - c_2(\dot{x}_2 - \dot{x}_1) + f_2(t) \end{aligned} \tag{3.19}$$

where it is assumed that the damping force is proportional to the velocity.

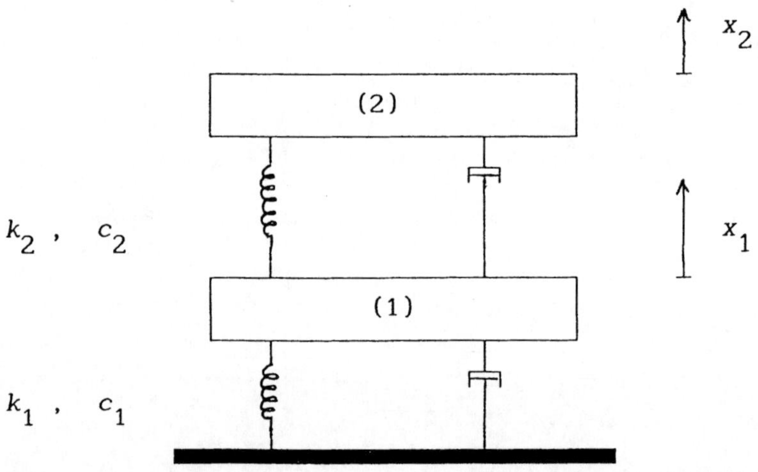

Figure 3.12

In matrix notation the system (3.19) may be written

$$M\ddot{x} + C\dot{x} + Kx = F$$

where

$$M = \begin{bmatrix} m_1 & 0 \\ 0 & m_2 \end{bmatrix}, \quad C = \begin{bmatrix} c_1 + c_2 & -c_2 \\ -c_2 & c_2 \end{bmatrix}, \quad K = \begin{bmatrix} k_1 - k_2 & -k_2 \\ -k_2 & k_2 \end{bmatrix}$$

are called the *mass*, *damping* and *stiffness* matrices, respectively and the external force is given by the vector function

$$F(t) = [f_1(t)\ \ f_2(t)]^T.$$

The inverse problem of determining the matrices M, C, K is of considerable importance in the structural dynamics industry and serious investments in hardware, software and personnel have been made to solve it effectively. The methods are based on judicious choices of the forcing function $F(t)$ ("shaking" the structure) and careful measurement and analysis of the response x.

3.5 Diffusion Coefficients

In the steady state, diffusive phenomena are modeled by the partial differential equation

$$-\nabla \cdot (a\nabla u) = f \tag{3.20}$$

where the function a is a diffusion coefficient and f represents a source term. For example, $u(x)$ may represent the temperature at point x in a three-dimensional body, $a(x)$ would be a variable thermal diffusivity coefficient and f represents external heating. The partial differential equation (3.20) also models underground steady state aquifers in which the coefficient $a(x)$ is the aquifer transmissivity coefficient and $u(x)$ represent the hydraulic head.

We will consider a one-dimensional version of (3.20) namely,

$$-(a(x)u'(x))' = f(x), \quad 0 < x < 1 \tag{3.21}$$

along with the boundary conditions

$$a(0)u'(0) = b_0, \quad a(1)u'(1) = b_1.$$

This boundary value problem may be thought of as a model for one-dimensional steady state temperature distributions in an inhomogeneous bar with source terms and specification of heat flux at the ends of the bar. If we are willing to think of x as representing time, u as representing position, a as representing a variable mass and f as representing an external force, then the boundary value problem models the motion of an object with momentum specifications at times $x = 0$ and $x = 1$.

The inverse problem we wish to consider briefly is that of estimating the coefficient $a(x)$ based on knowledge of u and f.

Exercise 3.19: Suppose that

$$f(x) = \begin{cases} -4x + 1, & x \in [0, 1/2] \\ 0, & x \in [1/2, 1] \end{cases} \quad \text{and} \quad u(x) = \begin{cases} x^2 - x + 5/4, & x \in [0, 1/2] \\ 1, & x \in [1/2, 1] \end{cases}.$$

Show that for any differentiable function φ on $[1/2, 1]$ with $\varphi(1) = 0$ and $\varphi'(1/2) = 1$ the coefficient

$$a(x) = \begin{cases} x, & x \in [0, 1/2] \\ \varphi(x), & x \in [1/2, 1] \end{cases}$$

satisfies (3.21). Therefore, in this case the inverse problem has infinitely many solutions. □

Exercise 3.20: Let $f(x) = -1$ for $x \in [0, 1]$ and let $b_0 = 0$, $b_1 = 1$. Show that if $u(x) = x$, then $a(x) = x$ satisfies (3.21). Show that if u is perturbed to

$$u_\epsilon(x) = \epsilon \sin(x/\epsilon^2) + x$$

then (3.21) is satisfied with

$$a_\epsilon(x) = \epsilon x / (\epsilon + \cos(x/\epsilon^2)).$$

Observe that, as $\epsilon \to 0$, $u_\epsilon \to u$ uniformly, but $a_\epsilon \to 0 \neq a$. Hence the coefficient identification problem is unstable with respect to perturbations in the observation u. □

Exercise 3.21: Let $u_a(x)$ be the solution of (3.21) with coefficient a and $u_b(x)$ be the solution of (3.21) when the coefficient is b. Show that

$$a(x) - b(x) = b_0(u_b'(x) - u_a'(x))(b_0 - \int_0^x f(s)ds)/(u_a'(x)u_b'(x)).$$

In what sense should deviations in u be measured in order that small deviations in u will lead to small deviations in the coefficient a? □

Finally, we indicate very briefly a possible approach to estimating the coefficient $a(x)$. Suppose, for simplicity, that $b_0 = 0$. Given u and f satisfying (3.21), we may attempt to identify a by assuming as an approximation a piecewise linear form

$$a_h = \sum_{j=0}^{N} c_j^{(h)} \varphi_j \tag{3.22}$$

where $h = 1/N$, for a given positive integer N and $\{\varphi_j\}_{j=0}^N$ are the familiar "tent" functions, that is, φ_j is continuous and is linear on each interval $[ih, (i+1)h]$ and $\varphi_j(ih) = 0$ for $i \neq j$, $\varphi_j(jh) = 1$. These functions are pictured in Figure 3.13.

Suppose that u_h is the solution of (3.21) using the coefficient a_h from (3.22). In Exercise 2.20 we noted that a_h will be near a if u_h' is near u', therefore, we will attempt to choose the coefficients $c_j^{(h)}$ so that the difference $u_h' - u'$ is small in some sense. Now,

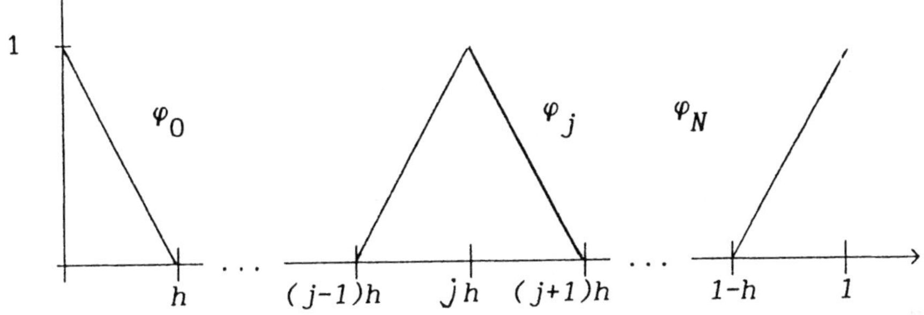

Figure 3.13

$$u_h'(x) = (-\int_0^x f(s)ds)/a_h(x)$$

and hence

$$u_h'(x_j) = (-\int_0^{x_j} f(s)ds)/c_j^{(h)}$$

where $x_j = jh$. A least squares approach for choosing the coefficients $c_j^{(h)}$ could then be based on minimizing the quantity

$$\sum_{j=0}^{N}(-\int_0^{x_j} f(s)ds/c_j^{(h)} - u'(x_j))^2. \tag{3.23}$$

In practice, values of f and u will be subject to measurement error. Such errors in f will not cause serious instability in computing the integrals in (3.23), however, as we have observed repeatedly above, the numerical calculation of the derivatives of u called for (3.23) can, and will, lead to numerical instability.

3.6 Forcing Terms

The nonhomogeneous term in a differential equation typically represents an outside influence, such as an external force, acting on a system. It is common to refer to such a term as a *forcing* or *source* term. The pressure gauge example in Chapter 2 involved finding such a forcing term, which depended on time but not space, given an observed response of the dynamical system. However, very often the forcing term is distributed, that is, it depends both on time and space. The determination of a distributed forcing term is then akin to the problem of identifying distributed coefficients in differential equations.

In this brief section we pose the problem of finding a source term, $f(x,t)$, in the one-dimensional heat equation. That is, we deal with the partial differential equation

$$u_t = u_{xx} + f(x,t), \quad 0 < x < \pi, \quad t > 0. \tag{3.24}$$

The term $f(x,t)$ represents a rate of production of thermal energy per unit time per unit length. For simplicity, we assume homogeneous boundary and initial conditions (see Exercise 3.22):

$$\begin{aligned} u(x,0) &= 0, \quad 0 < x < \pi \\ u(0,t) &= u(\pi,t) = 0. \end{aligned} \tag{3.25}$$

The problem we pose is the determination of the source term $f(x,t)$ from temperature measurements $u(a,t)$ at an interior site a, where $0 < a < \pi$.

To see the relationship between the source term and the interior temperatures, we will work formally with Fourier series. Suppose that $f(x,t)$ has a Fourier sine expansion

$$f(x,t) = \sum_{n=1}^{\infty} f_n(t) \sin nx$$

where the coefficients are given by

$$f_n(t) = \frac{2}{\pi} \int_0^{\pi} f(x,t) \sin nx \, dx. \tag{3.26}$$

Assume a like expansion for $u(x,t)$,

$$u(x,t) = \sum_{n=1}^{\infty} u_n(t) \sin nx, \tag{3.27}$$

we find on substituting into (3.24) and using the initial condition in (3.25) that the coefficients $u_n(t)$ satisfy the nonhomogeneous linear initial value problems

$$\begin{aligned} u_n' + n^2 u_n &= f_n \\ u_n(0) &= 0. \end{aligned} \tag{3.28}$$

A routine application of Laplace transforms solves (3.28) yielding

$$u_n(t) = \int_0^t e^{-n^2(t-\tau)} f_n(\tau) d\tau. \tag{3.29}$$

We note at this point that, for each given positive integer n, the coefficient $u_n(t)$ is purely *temporal*, however the entire collection of coefficients $\{u_n(t)\}$ contains all the *spatial* information in $u(x,t)$ as reflected in (3.27). From (3.29) we then see that the action of the exponential term will severely damp details in $f_n(t)$ and hence the recovery of the source term f from information in u will generally be a very difficult task (see Exercise 3.23).

To complete the analysis of the relationship between f and u, substitute (3.26) into (3.29) and use the result in (3.27). Interchanging the order of summation and integration and substituting the interior point $x = a$ we find

$$u(a,t) = \int_0^t \int_0^\pi k(s, t-\tau) f(s,\tau) ds\, d\tau$$

where

$$k(s,z) = \frac{2}{\pi} \sum_{n=1}^{\infty} e^{-n^2 z} \sin na \sin ns.$$

Again we see an inverse problem phrased in terms of an integral equation of the first kind.

Exercise 3.22: Show that if u satisfies heat equation with source term f, i.e., equation (3.24) and homogeneous boundary conditions (3.25) and if v satisfies the heat equation without a source term and nonhomogeneous boundary conditions, then $u+v$ satisfies the heat equation with source term f and nonhomogeneous boundary conditions. □

Exercise 3.23: Suppose that $u(x,t) = n^{-3/2}(2 - e^{-n^2 t}) \sin nx$ and $f(x) = 2\sqrt{n} \sin nx$. Show that u and f satisfy (3.24) and that $u(0,t) = u(\pi,t) = 0$. Also show that $u(x,t) \to 0$ uniformly on $[0,\pi] \times (0,\infty)$ as $n \to \infty$, but $\max \mid f(x) \mid \to \infty$ as $n \to \infty$.
□

3.7 Bibliographic Notes

The first example, relating to laws of growth, was chosen for its simplicity and familiarity. Other inverse problems related to coefficient identification in laws of growth could, of course, be posed. [Sa] is an excellent introduction to mathematical laws of growth. For a fuller discussion of forward differences for approximating derivatives of functions with errors see [Gro 4]. The direct hydraulic problem of section 2 is treated in some elementary texts; the inverse problem was inspired by a visit to the sacrificial cenote at Chichén Itzá (see [Gro 6]). [A] is an excellent reference on compartmental analysis (see also [Bel], [God] and [Mi]). For more on inverse problems in vibration, see [Gl].

The section on diffusion coefficients is based on [F] (see also [J]). For a related inverse problem involving the identification of an aquifer transmissity coefficient, see [AD]. [BK] is a very good source on identification of distributed parameters. For related information on numerical methods for nonlinear least squares problems, see [DS]. The section on forcing terms is based on chapter 19 of [Ca]. Exercise 3.23 comes from [Ru].

4 Mathematical Background for Inverse Problems

One of the most pervasive ideas in
mathematics is that of *geometrization*
D. Ruelle, Chance and Chaos

The inverse problems introduced in the previous chapters involve finding unknown functions (including functions defined on finite sets, that is, vectors or matrices) given other functions which are transformed versions of the desired functions. It is therefore no surprise that an appropriate mathematical framework for the analysis of inverse problems turns out to be the theory of function spaces and the fundamentals of the theory of operators on such spaces. In this chapter we sketch the basics of the theory of linear spaces, particularly Hilbert space, and the corresponding operator theory. Our aim is to introduce terminology, notation and basic concepts. For the most part, proofs are omitted; they can be found in the usual sources. The issues of existence, uniqueness and stability of solutions of linear inverse problems are addressed within the context of ill-posed operator equations and generalized inverses in the third section.

4.1 A Function Space Précis

We take for granted that the reader is familiar with the definition and basic properties of a linear space. Generally, we will deal with real normed linear spaces, that is, linear spaces for which the field of scalars is the field of real numbers and which are endowed with a real-valued function $||\cdot||$ having the properties of length, that is,

$$||f|| \geq 0 \text{ and } ||f|| = 0 \Leftrightarrow f = 0$$

$$||\alpha f|| = |\alpha|\, ||f||,\ \alpha \in \mathbf{R}$$

and

$$||f + g|| \leq ||f|| + ||g||.$$

The space $C[a,b]$ of continuous real-valued functions on the interval $[a,b]$ with the norm

$$||f||_\infty = \max_{a \leq x \leq b} |f(x)|$$

is a familiar example of a normed linear space.

Sometimes the norm $|| \cdot ||$ on a linear space H is engendered by an *inner product*, that is, a real-valued, symmetric, definite, bilinear form $< \cdot, \cdot >$. In detail, $< \cdot, \cdot >$: $H \times H \rightarrow \mathbf{R}$ is an inner product on H if

$$< f, f > \geq 0 \text{ and } < f, f > = 0 \Leftrightarrow f = 0$$

$$< f, g > = < g, f >$$

and

$$< \alpha f + \beta g, h > = \alpha < f, h > + \beta < g, h >, \quad \alpha, \beta \in \mathbf{R}.$$

The space $C[a, b]$ with inner product

$$< f, g > = \int_a^b f(t) g(t) dt \tag{4.1}$$

and the space $C[a, b]$ of continuously differentiable functions on $[a, b]$ with inner product

$$< f, g > = \int_a^b f'(t) g'(t) dt + f(a) g(a),$$

are typical examples of inner product spaces.

Each inner product $< \cdot, \cdot >$ gives rise to a norm $|| \cdot ||$ defined by

$$||f|| = \sqrt{< f, f >}.$$

Also, since the quadratic $< f + tg, f + tg >$ in $t \in \mathbf{R}$ is always nonnegative, its discriminant must be nonpositive. This leads to the following important result, known as the *Cauchy-Schwarz inequality*

$$| < f, g > | \leq ||f|| \, ||g||. \tag{4.2}$$

Exercise 4.1: Show that if $|| \cdot ||$ is a norm generated by an inner product, then it satisfies the *parallelogram law*

$$||f + g||^2 + ||f - g||^2 = 2||f||^2 + 2||g||^2.$$

Show that $C[a, b]$ with the norm $|| \cdot ||_\infty$ is not an inner product space. □

A sequence $\{f_n\}$ in a normed linear space is said to *converge* (strongly) to f, denoted $f_n \rightarrow f$, if

$$||f_n - f|| \rightarrow 0 \text{ as } n \rightarrow \infty.$$

If $< f_n, g > \rightarrow < f, g >$ for each g in an inner product space, we say that $\{f_n\}$ *converges weakly* to f. Weak convergence is denoted $f_n \rightharpoonup f$. The Cauchy-Schwarz inequality shows that strong convergence implies weak convergence. The identity

$$||f_n - f||^2 = ||f_n||^2 - 2 < f_n, f > + ||f||^2$$

shows that $f_n \rightharpoonup f$ and $||f_n|| \to ||f||$ is equivalent to $f_n \to f$.

A normed linear space is called *complete* if each Cauchy sequence in the space converges, that is, if $||f_m - f_n|| \to 0$ as $m, n \to \infty$ implies there is an f in the space with $f_n \to f$. By a standard abstract construction (equivalence classes of Cauchy sequences) any normed linear space can be completed. The *completion* of a normed linear space is the smallest normed linear space containing the given space. A *Hilbert space* is a complete inner product space. The most familiar Hilbert space is the space $L^2[a, b]$ of all (equivalence classes of) Lebesgue measurable functions whose squares have finite integrals over $[a, b]$. $L^2[a, b]$ is the completion of $C[a, b]$ with respect to the inner product (4.1). The *Sobolev space* $H^m[a, b]$ is another important Hilbert space. $H^m[a, b]$ is the completion of the space of functions having m continuous derivatives on $[a, b]$ for which the norm

$$||f||_m = (\sum_{j=0}^{m} ||f^{(j)}||_0^2)^{1/2}$$

is finite ($||\cdot||_0$ is the L^2-norm). The inner product associated with this norm is

$$< f, g >_m = \sum_{j=0}^{m} < f^{(j)}, g^{(j)} >$$

where $< \cdot, \cdot >$ is the L^2-inner product. Note that $L^2[a, b] = H^0[a, b]$.

A subset S of a normed linear space is called (strongly) *closed* if $\{f_n\} \subseteq S$ and $f_n \to f$ implies $f \in S$. If the same condition holds in an inner product space, with strong convergence replaced by weak convergence, then S is called *weakly closed*. Hilbert spaces enjoy a kind of weak Bolzano-Weierstrass property: *every (norm) bounded sequence in a Hilbert space has a weakly convergent subsequence*. From this follows a very important approximation property of weakly closed sets:

> *If C is a weakly closed set in a Hilbert space H and $x_0 \in H$, then there is a $y \in C$ with*
>
> $$||y - x_0|| = \inf\{||w - x_0|| : w \in C\}.$$
>
> *If, in addition, C is convex, then the proximal vector y is unique.*

To see why this is so, we may take $x_0 = 0$. Let $d = \inf\{||w|| : w \in C\}$ and choose a sequence $\{y_n\} \subseteq C$ with $||y_n|| \to d$. Extract a weakly convergent subsequence with $y_{n_k} \rightharpoonup y \in C$. Then

$$||y||^2 = \lim_{k\to\infty} < y, y_{n_k} > \leq ||y|| \lim_{k\to\infty} ||y_{n_k}|| = ||y||d$$

and hence $||y|| = d$, as required. If, in addition, C is convex and there is a vector $w \in C$ satisfying $||w|| = d$ and $w \neq y$, then, by Exercise 4.1,

$$||(w+y)/2||^2 = -\frac{1}{4}||w-y||^2 + \frac{1}{2}(||w||^2 + ||y||^2) < d^2,$$

a contradiction.

We remark that if a subset C of a Hilbert space is strongly closed and convex, then (as a consequence of the Hahn-Banach theorem) it is also weakly closed.

Two vectors, f and g, in an inner product space H are called *orthogonal* if $< f, g > = 0$. The *orthogonal complement* of a set $S \subseteq H$ is the closed subspace

$$S^\perp = \{g \in H : < f, g > = 0 \text{ for all} f \in S\}.$$

A closed subspace S of a Hilbert space H engenders an *orthogonal decomposition*, $H = S \oplus S^\perp$, meaning that each $f \in H$ can be written in a unique way as $f = f_1 + f_2$, where $f_1 \in S$ and $f_2 \in S^\perp$. The vector f_1 is the *projection* of f onto S, that is, the unique vector in S satisfying

$$||f - f_1|| = \inf\{||f - g|| : g \in S\}.$$

It is then easy to see that $f_2 = f - f_1 \in S^\perp$.

A set of vectors $\{f_\alpha\}_{\alpha \in A} \subseteq H$ is said to be *orthogonal* if $< f_\alpha, f_\beta > = 0$ for $\alpha \neq \beta$. If, moreover, $||f_\alpha|| = 1$ for each $\alpha \in A$, the set is called *orthonormal.* If $\{f_\alpha\}_{\alpha \in A}$ is an orthonormal set, then for any $f \in H$ only countably many of the numbers $< f, f_\alpha >$ is nonzero and *Bessel's inequality* holds:

$$\sum_{\alpha \in A} | < f, f_\alpha > |^2 \leq ||f||^2. \tag{4.3}$$

A *complete orthonormal set* is an orthonormal set $\{f_\alpha\}_{\alpha \in A}$ with the property that $< f, f_\alpha > = 0$ for all $\alpha \in A$ only if $f = 0$. If $\{f_\alpha\}_{\alpha \in A}$ is a complete orthonormal set, then each $f \in H$ has a unique *Fourier expansion*

$$f = \sum_{\alpha \in A} < f, f_\alpha > f_\alpha$$

and *Parseval's identity* is valid:

$$||f||^2 = \sum_{\alpha \in A} | < f, f_\alpha > |^2.$$

A Hilbert space is *separable* if it contains a countable complete orthonormal set. The set $\{\sin n\pi x\}_{n=1}^{\infty}$ is a complete orthonormal set for the separable Hilbert space $L^2[0,1]$.

We close this section with a few words on compactness. A subset S of a normed linear space is called *compact* if each sequence in S has a subsequence which converges to an element of S. For example, a subset of $\mathbf{R}^n$ is compact if and only if it is closed and bounded. The *Arzela-Ascoli Theorem* characterizes compact sets in $C[a,b]$. It says that a closed set S in $C[a,b]$ is compact if and only if it is bounded and *equicontinuous*, that is, given $\epsilon > 0$ there is a $\delta > 0$ such that for any $f \in S$

$$t, x \in [a, b] \text{ and } |x - t| < \delta \Rightarrow |f(x) - f(t)| < \epsilon.$$

A special case of the *Sobolev Imbedding Theorem* asserts that a set of the form

$$S = \{f \in H^1[0,1] : \|f\|_1 \leq C\}$$

in $H^1[0,1]$ is compact in $C[0,1]$. We sketch why this is so. Any function $f \in H^1[0,1]$ is absolutely continuous [Mar 2] and we can write

$$f(x) = f(t) + \int_t^x f'(s)ds \tag{4.4}$$

and hence (integrating in t):

$$\begin{aligned} f(x) &= \int_0^1 f(t)dt + \int_t^x f'(s)dsdt \\ &= \int_0^1 f(t)dt + \int_0^x sf'(s)ds. \end{aligned}$$

Therefore, by the Cauchy-Schwarz inequality (applied first on $L^2[0,1]$ and then on $\mathbf{R}^2$):

$$\begin{aligned} |f(x)| &\leq |\int_0^1 f(t)dt| + \int_0^1 s|f'(s)|ds \\ &\leq \|f\|_0 + \|f'\|_0 \leq \sqrt{2}\|f\|_1 \leq \sqrt{2}C. \end{aligned}$$

Therefore, $\|f\|_\infty \leq \sqrt{2}C$, for all $f \in S$, that is, S is bounded in $C[0,1]$. Also, from (4.4), we have (thanks to Cauchy-Schwarz)

$$\begin{aligned} |f(x) - f(t)| &\leq |\int_t^x f'(s)ds| \\ &\leq |x-t|^{1/2}\|f'\|_0 \\ &\leq |x-t|^{1/2}\|f\|_1 \leq C|x-t|^{1/2} \end{aligned}$$

and hence the functions in S are equicontinuous. Since S is closed in $C[0,1]$, it follows from the Arzela-Ascoli theorem that S is compact in $C[0,1]$.

4.2 Some Operator Theory

The term *operator* is used to describe a function defined on a subset of a linear space and taking values in another linear space. For example, the evaluation operator E_0 defined on $C[0,1]$ by

$$E_0 f = f(0)$$

is an operator from $C[0,1]$ to $\mathbf{R}$, and the differentiation operator D defined on the subset $\mathcal{D}$ of $C[0,1]$ consisting of continuously differentiable functions by

$$Df = f'$$

is an operator from $\mathcal{D}$ into $C[0,1]$.

Suppose $\mathcal{D}$ is a linear subspace of a linear space. An operator T defined on $\mathcal{D}$ and taking values in another linear space is called a *linear operator* if

$$T(\alpha f + \beta g) = \alpha T f + \beta T g$$

for any vectors f and g and any scalars α and β. Each of the operators E_0 and D defined above is linear as are the operators on finite dimensional spaces represented by matrices (relative to given bases). If $k(\cdot,\cdot) \in C([c,d] \times [a,b])$, then the integral operator

$$(Kf)(s) = \int_a^b k(s,t)f(t)dt, \quad s \in [c,d]$$

is a linear operator from the space $C[a,b]$ into the space $C[c,d]$. A linear operator defined on a linear subspace of a linear space and taking values in the scalar field is called a *linear functional*. For example, the evaluation operator E_0 defined above is a linear functional on the space $C[0,1]$.

A linear operator T defined on a linear subspace $\mathcal{D}(T)$ of a normed linear space and taking values in another normed linear space is called *bounded* if there is a number M satisfying

$$||Tf|| \leq M||f|| \tag{4.5}$$

for all $f \in \mathcal{D}(T)$ (note that the same symbol has been used for the norm in both the domain and range space). If T is a bounded linear operator defined everywhere on a normed linear space, then the smallest value of M satisfying (4.5) is called the *norm*, $||T||$, of T, that is,

$$||T|| = \sup_{f \neq 0} \frac{||Tf||}{||f||}. \tag{4.6}$$

We will generally reserve the term *bounded linear operator* for an everywhere defined linear operator with finite norm. The bounded linear operators on a normed linear space are precisely the everywhere defined continuous linear operators and the space of all bounded linear operators endowed with the norm (4.6) is itself a normed linear space. Bounded linear functionals on a Hilbert space have a particularly simple form. The *Riesz Representation Theorem* states that for each bounded linear functional G on a Hilbert space H there is a unique element $g \in H$ satisfying

$$G(f) = < f, g >$$

for all $f \in H$. Moreover, $||G|| = ||g||$.

Integral operators provide typical examples of bounded linear operators on Hilbert space, while differential operators are the usual exemplars of unbounded linear operators. For example, if $k(\cdot,\cdot) \in L^2([0,1] \times [0,1])$, then the integral operator on $L^2[0,1]$ defined by

$$(Kf)(s) = \int_0^1 k(s,t)f(t)dt \tag{4.7}$$

satisfies, by the Cauchy-Schwarz inequality

$$||Kf||_0 \leq M||f||_0$$

where

$$M = \left\{ \int_0^1 \int_0^1 |k(s,t)|^2 dt ds \right\}^{1/2}.$$

On the other hand, if $\mathcal{D}(T) = C^1[0,1] \subseteq L^2[0,1]$ and $T : \mathcal{D}(T) \to L^2[0,1]$ is defined by

$$Tf = f',$$

then $f_n(x) = \sin n\pi x \in \mathcal{D}(T)$, $||f_n||_0 = 1/\sqrt{2}$ and

$$||Tf_n||_0 = n\pi/\sqrt{2} \to \infty \text{ as } n \to \infty$$

and hence T is unbounded.

If $T : H_1 \to H_2$ is a bounded linear operator from a Hilbert space H_1 into a Hilbert space H_2, then the *adjoint* of T is the operator $T^* : H_2 \to H_1$ satisfying

$$< Tf, g > = < f, T^* g >$$

for all $f \in H_1$ and $g \in H_2$. If $H_1 = H_2$ and $T = T^*$, then T is called *self-adjoint*.

Exercise 4.2: Show that if $T : H_1 \to H_2$ is a bounded linear operator, then the adjoint T^* exists, is a bounded linear operator and satisfies:

$$||T^*|| = ||T||, \quad T^{**} = T, \quad (TS)^* = S^* T^*. \ \square$$

Exercise 4.3: Suppose that $k(\cdot,\cdot) \in L^2([0,1] \times [0,1])$. Show that if K is the operator defined on the real space $L^2[0,1]$ by

$$(Kf)(s) = \int_0^1 k(s,t)f(t)dt,$$

then

$$(K^* g)(t) = \int_0^1 k(s,t)g(s)ds.$$

In particular, if the kernel is *symmetric*, that is, $k(s,t) = k(t,s)$, then the operator K is self-adjoint. Also show that

$$(K^*Kf)(s) = \int_0^1 \tilde{k}(s,t)f(t)dt$$

where

$$\tilde{k}(s,t) = \int_0^1 k(u,s)k(u,t)du. \quad \Box$$

Exercise 4.4: The *nullspace* $N(T)$ of an operator T is the set

$$N(T) = \{x \in \mathcal{D}(T) : Tx = 0\}.$$

Show that the nullspace of a bounded linear operator $T : H_1 \rightarrow H_2$ is a closed subspace, that $N(T^*) = R(T)^\perp$, where

$$R(T) = \{y : Tx = y, \ \text{some } x \in H_1\}$$

is the *range* of T, and that $N(T)^\perp = \overline{R(T^*)}$ (the bar indicates strong closures). □

A bounded linear operator K from a normed linear space X into a normed linear space Y is called *compact* if for each bounded set B in X, the set $K(B)$ has compact closure in Y. For example, the Sobolev imbedding theorem, a special case of which was discussed in the previous section, guarantees that the imbedding operator $K : H^1[0,1] \rightarrow C[0,1]$ given by $Kf = f$ is a compact operator. It follows from the definition that bounded linear operators having finite dimensional range are compact. Also, any linear operator on a complete normed linear space which is the limit, in operator norm, of a sequence of compact operators is itself a compact operator, i.e., the space of compact operators is closed in the operator norm.

Exercise 4.5: Suppose $k(\cdot,\cdot) \in C([0,1] \times [0,1])$. Show that the operator $K : C[0,1] \rightarrow C[0,1]$ defined by (4.7) is compact. □

If $T : H \rightarrow H$ is a bounded linear operator on a Hilbert space, then the *spectrum* of T, $\sigma(T)$, is the set of complex numbers λ for which the operator $T - \lambda I$ has no bounded inverse (I denotes the identity operator on H). The spectrum of a bounded linear operator T is a compact subset of the complex plane. If T is bounded and self-adjoint, then $\sigma(T) \subseteq \mathbf{R}$ and the norm of T is equal to its *spectral radius*:

$$\|T\| = \max\{|\lambda| : \lambda \in \sigma(T)\}.$$

A number λ is called an *eigenvalue* of T if

$$Tf = \lambda f$$

for some nonzero vector f (called an *eigenvector* associated with λ).

The spectral life of a compact self-adjoint linear operator T on a Hilbert space is fairly simple: $\sigma(T)$ is nonempty and consists only of real numbers, every nonzero member of $\sigma(T)$ is an eigenvalue of T and for each nonzero eigenvalue λ, the *eigenspace* $N(T - \lambda I)$ is finite dimensional. Moreover, the nonzero eigenvalues of T may be arranged in a sequence $|\lambda_1| > |\lambda_2| > \ldots$ which is finite if $R(T)$ is finite dimensional and which satisfies $\lambda_n \to 0$ if $R(T)$ is infinite dimensional. Finally, eigenvectors associated with distinct eigenvalues are orthogonal and the set of all eigenvectors associated with the nonzero eigenvalues is complete in $\overline{R(T)} = N(T)^{\perp}$.

The *spectral theorem* for a compact self-adjoint operator then allows us to arrange the eigenvalues in a sequence $\lambda_1, \lambda_2, \lambda_3, \ldots$ with $|\lambda_1| \geq |\lambda_2| \geq \ldots$ and to construct a corresponding sequence $v_1, v_2, v_3, \ldots$ of orthonormal eigenvectors such that every vector $w \in H$ has an eigenfunction expansion

$$w = Pw + \sum_{j=1}^{\infty} < w, v_j > v_j \tag{4.8}$$

where Pw is the orthogonal projection of w onto $N(T)$. It then follows that

$$Tw = \sum_{j=1}^{\infty} \lambda_j < w, v_j > v_j \tag{4.9}$$

(if $R(T)$ is finite dimensional, the sums above are finite).

The spectral representation (4.9) allows us to define certain functions of the operator T. If, for example, f is a continuous real-valued function defined on a closed interval containing $\sigma(T)$ we can define the self-adjoint operator $f(T)$ by

$$f(T)w = \sum_{j=1}^{\infty} f(\lambda_j) < w, v_j > v_j$$

The *spectral mapping theorem* then asserts that $\sigma(f(T)) = f(\sigma(T))$ and the spectral radius formula gives

$$\|f(T)\| = \max\{|f(\lambda)| : \lambda \in \sigma(T)\}.$$

Exercise 4.6: Let

$$k(s,t) = \begin{cases} s(1-t), & 0 \leq s < t \\ t(1-s), & t \leq s \leq 1. \end{cases}$$

Show that the integral operator K on $L^2[0,1]$ generated by $k(\cdot,\cdot)$ is a compact self-adjoint operator. Also, show that the eigenvalues of K are $\lambda_n = 1/(n\pi)^2$, $n = 1, 2, \ldots$ and that $v_n(x) = \sqrt{2} \sin n\pi x$ are corresponding orthonormal eigenfunctions. (Hint: See Exercise 2.1 relating to the hanging cable model.) □

A Fredholm integral equation of the second kind in $L^2[0,1]$,

$$-\lambda f(s)+\int_0^1 k(s,t)f(t)dt=g(s),$$

where λ is a given nonzero number, can be translated into an operator equation

$$(K-\lambda I)f=g \tag{4.10}$$

where K is the integral operator generated by $k(\cdot,\cdot)$. If $k(\cdot,\cdot)$ is symmetric and square integrable, then K is compact and self-adjoint and hence we may apply the spectral theory outlined above to the equation (4.10). In particular, we see that if λ is not an eigenvalue of K, then $(K-\lambda I)^{-1}$ is a bounded linear operator. Therefore, equation (4.10) has, for each $g \in L^2[0,1]$, a unique solution

$$f=(K-\lambda I)^{-1}g$$

in $L^2[0,1]$ which depends continuously on g. That is, (4.10) is a *well-posed* problem if λ is not an eigenvalue of K.

On the other hand, if λ is an eigenvalue of K all is not lost. In this case (4.10) has a solution if and only if

$$g \in R(K-\lambda I)=N(K-\lambda I)^{\perp}$$

that is, if and only if g is orthogonal to all eigenvectors associated with the eigenvalue λ. Assuming this to be the case, any function of the form

$$f=\sum_j{}' \frac{<g,v_j>}{\lambda_j-\lambda}v_j-\lambda^{-1}Pg+\psi \tag{4.11}$$

where ψ is any function in the eigenspace $N(K-\lambda I)$, Pg is the orthogonal projection of g onto $N(K)$ and the sum $\sum_j'$ indicates that terms satisfying $\lambda_j=\lambda$ are omitted, is a solution of (4.10) and all solutions of (4.10) have the form (4.11).

In short, if λ is an eigenvalue of K, then (4.10) has a solution only if g satisfies the additional condition that it is orthogonal to all eigenfunctions associated with λ. If this is the case, then (4.10) has infinitely many solutions of the form (4.11). Note, however, that there is a unique solution (4.11) having minimum norm, namely

$$f=\sum_j{}' \frac{<g,v_j>}{\lambda_j-\lambda}v_j-\lambda^{-1}Pg$$

and this solution depends continuously on g. In this sense, the Fredholm integral equation of the second kind (4.10) with self-adjoint compact operator K, is a well-posed problem regardless of the (nonzero) value λ.

In the first chapter we saw several examples of ill-posed integral equations of the first kind. We will now consider such equations in an abstract setting, that is, we consider an equation of the form

$$Kf = g \tag{4.12}$$

where $K : H_1 \to H_2$ is a compact, linear (but not necessarily self-adjoint) operator from a Hilbert space H_1 into a Hilbert space H_2. Our analysis hinges on the behavior of the compact self-adjoint operators $K^*K : H_1 \to H_1$ and $KK^* : H_2 \to H_2$. It is easy to see that these two operators have the same eigenvalues and that the nonzero eigenvalues are positive. Let $\lambda_1 \geq \lambda_2 \geq \ldots$ be an enumeration of these positive eigenvalues and let $v_1, v_2, \ldots$ be a sequence of associated orthonormal eigenvectors of K^*K. Then $\{v_1, v_2, \ldots\}$ is complete in $\overline{R(K^*K)} = N(K)^\perp$. Let $\mu_j = \sqrt{\lambda_j}$ and $u_j = \mu_j^{-1} K v_j$. Then

$$K^* u_j = \mu_j v_j \tag{4.13}$$

and

$$K v_j = \mu_j u_j. \tag{4.14}$$

Moreover,

$$KK^* u_j = \mu_j K v_j = \mu_j^2 u_j = \lambda_j u_j$$

and it is not hard to see that the orthonormal eigenvectors $\{u_j\}$ of KK^* form a complete orthonormal set for $\overline{R(KK^*)} = N(K^*)^\perp$. The system $\{v_j, u_j; \mu_j\}$ is called a *singular system* for the operator K and the numbers μ_j are called *singular values* of K. Any $f \in H_1$ has a representation

$$f = Pf + \sum_{j=1}^{\infty} < f, v_j > v_j$$

where P is the orthogonal projector of H_1 onto $N(K)$ and hence

$$Kf = \sum_{j=1}^{\infty} \mu_j < f, v_j > u_j \tag{4.15}$$

This representation of the operator K is called the *singular value decomposition* (SVD for short).

If the equation of the first kind (4.12) has a solution f, then $g \in R(K)$ and, by (4.14)

$$\lambda_j^{-1} |< g, u_j >|^2 = \lambda_j^{-1} |< Kf, \mu_j^{-1} K v_j >|^2 = |< f, v_j >|^2$$

and hence, by Bessel's inequality

$$\sum_{j=1}^{\infty} \lambda_j^{-1} |< g, u_j >|^2 = \sum_{j=1}^{\infty} |< f, v_j >|^2 \leq ||f||^2 < \infty.$$

Conversely, if $g \in \overline{R(K)} = N(K^*)^\perp$ and if

$$\sum_{j=1}^{\infty} \lambda_j^{-1} | < g, u_j > |^2 < \infty \tag{4.16}$$

then any function of the form

$$f = \sum_{j=1}^{\infty} \frac{< g, u_j >}{\mu_j} v_j + \varphi \tag{4.17}$$

where $\varphi \in N(K)$ is, by (4.15), a solution of (4.12). Our discussion can to some extent be summarized by stating *Picard's existence criterion*: equation (4.12) has a solution if and only if $g \in \overline{R(K)}$ and condition (4.16) holds.

Exercise 4.7: Suppose K is compact and $R(K)$ is not finite dimensional. Show that if (4.12) has a solution, then the solution with smallest norm is given by (4.17) with $\varphi = 0$. Show that the minimum norm solution does not depend continuously on g. (Hint: Consider perturbations to g of the form ϵu_n, where $\epsilon > 0$ is small and n is large.) □

Now some comments on unbounded operators. Suppose $T : \mathcal{D}(T) \subseteq H_1 \to H_2$ is a linear, but not necessarily bounded, operator, defined on a linear subspace $\mathcal{D}(T)$ of a Hilbert space H_1. The *graph* of T is the subspace

$$\mathcal{G}(T) = \{(f, Tf) : f \in \mathcal{D}(T)\}$$

of the product space $H_1 \times H_2$. The operator T is called *closed* if $\mathcal{G}(T)$ is a closed subspace of $H_1 \times H_2$. Note that this is equivalent to the condition

$$\{f_n\} \subseteq \mathcal{D}(T),\ f_n \to f \ \text{ and } \ Tf_n \to g \Rightarrow f \in \mathcal{D}(T) \ \text{ and } \ Tf = g.$$

A bounded linear operator is, of course, closed and a closed everywhere defined linear operator is bounded [RN].

If the domain of T, $\mathcal{D}(T)$, is dense in H_1 we can define an adjoint $T^* : \mathcal{D}(T^*) \subseteq H_2 \to H_1$, where $\mathcal{D}(T^*)$ is the space of all vectors $g \in H_2$ such that for some $h \in H_1$,

$$< Tf, g > = < f, h >$$

for all $f \in \mathcal{D}(T)$. This vector h is then uniquely determined and we define $T^* g = h$. If T is a closed densely defined linear operator, then T^* is likewise closed, densely defined and linear.

Exercise 4.8: Let $\mathcal{D}(T)$ be the space of all absolutely continuous functions f on $[0,1]$ satisfying $f(0) = 0$ and $f' \in L^2(0,1]$. Define $T : \mathcal{D}(T) \subseteq L^2[0,1] \to L^2[0,1]$ by $Tf = f'$. Show that T is closed. Find the adjoint T^* of T. Show that if $f \in \mathcal{D}(T^*T)$ and $T^*Tf = g$, then

$$f(s) = \int_0^1 k(s,t)g(t)dt$$

where

$$k(s,t) = \begin{cases} t, & 0 \le t \le s \\ s, & s \le t \le 1 \end{cases} \quad \square$$

4.3 Ill-Posed Operator Equations

A common abstract framework for inverse problems can be constructed in terms of operator equations of the first kind, that is, equations of the form

$$Kx = y \tag{4.18}$$

where $K : \mathcal{D}(K) \subseteq X \to Y$ is an operator defined on a subset $\mathcal{D}(K)$ of a normed linear space X and taking values in a normed linear space Y.

The equation (4.18) is well-posed if it has a unique solution $x \in \mathcal{D}(K)$ for each $y \in Y$ (that is, the inverse operator $K^{-1} : Y \to \mathcal{D}(K)$ exists) and this unique solution depends continuously on y (that is, K^{-1} is continuous). If (4.18) is well-posed, then x is *stable* with respect to small changes in y. It is the stability issue that is of primary concern when attempting to solve (4.18) because, as we saw in many of the examples in the previous chapters, in practical circumstances y is often a measured quantity and therefore is subject to observational errors. Stability then simply means that small errors in y will lead to small errors in the solution x. A classical theorem of Tikhonov, which is now a standard exercise in most topology texts, gives a simple stability result when K is restricted to compact sets. Tikhonov's theorem states:

> *If $K : \mathcal{D}(K) \to Y$ is a continuous one-to-one operator and $C \subseteq \mathcal{D}(K)$ is a compact set, then the inverse operator $(K|C)^{-1}$ is continuous ($K|C$ denotes the restriction of K to C).*

Exercise 4.9: Prove the theorem above. $\square$

To see how this result could apply to inverse problems, consider the hanging cable model of Chapter 1:

$$\int_0^1 k(s,t)x(t)dt = y(s)$$

where

$$k(s,t) = \begin{cases} s(1-t)/T, & 0 \le s \le t \\ t(1-s)/T, & t \le s \le 1 \end{cases},$$

$x(t)$ represents the unknown density distribution of the hanging cable and $y(s)$ is the observed sag of the cable at position s. If we consider the operator as acting on the space $C[0,1]$, and if we are willing to consider only densities satisfying an a priori bound of the type

$$||x||_1^2 = ||x||_0^2 + ||x'||_0^2 \leq C$$

(note that this would disallow the densities x_ϵ specified in Exercise 2.1), then we are dealing (by way of the Sobolev imbedding theorem) with a compact set of densities in $C[0,1]$. Tikhonov's theorem then guarantees that the inverse of the operator K restricted to this class is continuous, that is, the inverse problem is stable for this class of densities.

We now take up linear Fredholm integral equations of the first kind in a Hilbert space setting. Such equations may be phrased abstractly in the form

$$Kx = y \tag{4.19}$$

where $K : H_1 \to H_2$ is a bounded linear operator on a real Hilbert space H_1, taking values in a real Hilbert space H_2. Typically, these Hilbert spaces will be spaces of square integrable functions and the kernel will be a square integrable function of two variables, giving rise to a compact operator.

A solution x of (4.19) exists if and only if $y \in R(K)$. Since K is linear, $R(K)$ is a subspace of H_2, however, it generally does not exhaust H_2, as we have seen in many of the examples of Chapter 2. Therefore, a traditional solution of (4.19) will exist only for a restricted class of functions y. If we are willing to broaden our notion of solution, we may enlarge the class of functions y for which a type of *generalized* solution exists to a dense subspace of functions in H_2. This is accomplished by introducing the idea of a *least squares solution*. A function $x \in H_1$ is called a least squares solution of (4.19) if

$$||Kx - y|| = \inf\{||Ku - y|| : u \in H_1\}.$$

This is equivalent to saying that $Py \in R(K)$, where P is the orthogonal projector of H_2 onto $\overline{R(K)}$, the closure of the range of K. Now, $Py \in R(K)$ if and only if

$$y = Py + (I - P)y \in R(K) + R(K)^\perp. \tag{4.20}$$

Therefore, a least squares solution exists if and only if y lies in the dense subspace $R(K) + R(K)^\perp$ of H_2. By extending the notion of solution to the idea of least squares solution, we have guaranteed the existence of a generalized, i.e., least squares, solution of (4.19) for all y in a dense subspace of H_2.

In taking up the issue of uniqueness, we note that (4.20) is equivalent to the condition

$$Kx - y \in R(K)^\perp = N(K^*),$$

that is,

$$K^*Kx = K^*y, \tag{4.21}$$

where K^* is the adjoint of K. From (4.21) we see that there is a unique least squares solution if and only if

$$\{0\} = N(K^*K) = N(K),$$

and that the set of all least squares solutions is closed and convex. Therefore, *there is a unique least squares solution of smallest norm*, and it is this solution that we will adopt as our generalized solution of (4.19). The mapping $K^\dagger$ that associates with a given

$$y \in \mathcal{D}(K^\dagger) = R(K) + R(K)^\perp$$

the unique least squares solution having smallest norm, $K^\dagger y$, is called the *Moore-Penrose generalized inverse* of K.

Exercise 4.10: Suppose $y \in \mathcal{D}(K^\dagger)$. Show that $K^\dagger y$ is the unique least squares solution in $N(K)^\perp$ and that the set of all least squares solutions may be represented as $K^\dagger y + N(K)$. Also show that if $\bar{K}$ represents the operator K restricted to $N(K)^\perp$, then for any $y \in \mathcal{D}(K^\dagger)$, $K^\dagger y = \bar{K}^{-1}Py$, where P is the orthogonal projector of H_2 onto $\overline{R(K)}$. □

In our scheme $K^\dagger$ is then the mechanism which provides a unique (least squares) solution of (4.19) for any $y \in \mathcal{D}(K^\dagger)$. In this sense, $K^\dagger$ settles the issues of existence and uniqueness for generalized solutions of (4.19). The big issue remains. Namely, in order for (4.19) to be well-posed in the sense of Hadamard for generalized solutions it is necessary that $K^\dagger$ be continuous. The next result, which summarizes the basic properties of $K^\dagger$,shows precisely when this is the case.

> $K^\dagger : \mathcal{D}(K^\dagger) \to H_1$ *is a closed densely defined linear operator which is bounded if and only if* $R(K)$ *is closed.*

To see this, note first that $\mathcal{D}(K^\dagger) = R(K) + R(K)^\perp$ is evidently dense in H_2. The linearity of $K^\dagger$ follows easily from (4.21) and Exercise 4.10. To see that $K^\dagger$ is closed, note that if

$$\{y_n\} \subseteq \mathcal{D}(K^\dagger), \quad y_n \to y \text{ and } K^\dagger y_n \to x,$$

then $x \in N(K)^\perp$, since $K^\dagger y_n \in N(K)^\perp$, and that

$$K^* y_n = K^*KK^\dagger y_n \to K^*Kx.$$

But $K^*y_n \to K^*y$ and hence $K^*Kx = K^*y$, that is, x is a least squares solution lying in $N(K)^\perp$. Therefore we find $y \in \mathcal{D}(K^\dagger)$ and $K^\dagger y = x$, i.e., $K^\dagger$ is closed. Suppose now that $R(K)$ is closed, then $\mathcal{D}(K^\dagger) = H_2$ and $K^\dagger$ is a closed everywhere defined linear operator, therefore $K^\dagger$ is bounded. On the other hand, if $K^\dagger$ is bounded and $Kx_n \to y$, where $\{x_n\} \subseteq N(K)^\perp$, then

$$x_n = K^\dagger K x_n \to K^\dagger y \text{ and } Kx_n \to KK^\dagger y.$$

Therefore $y = KK^{\dagger}y \in R(K)$, and $R(K)$ is closed.

Exercise 4.11: Let $H_1 = H_2 = L^2[0,1]$ and define $K : H_1 \to H_2$ by

$$(Kx)(s) = \int_0^s x(t)dt.$$

Show that

$$R(K) = \{y \in L^2[0,1] : \ y \text{ is absolutely continuous}, \ y' \in L^2[0,1] \ \text{and} \ y(0) = 0\},$$

and that $K^{\dagger}y = y'$ if $y \in R(K)$. □

Exercise 4.12: The definition of $K^{\dagger}$ given above for a bounded linear operator K extends naturally to the case when K is a closed densely defined linear operator. Provide the details. Let $H_1 = H_2 = L^2[0,1]$ and let

$$\mathcal{D}(T) = \{x \in H_1 : \ x \text{ is absolutely continuous}, \ x' \in H_1, \ x(0) = x(1) = 0\},$$

and define $T : \mathcal{D}(T) \to H_2$ by $Tx = x'$. Show that $\mathcal{D}(T^{\dagger}) = H_2$ and that

$$(T^{\dagger}y)(t) = \int_0^t y(s)ds - t\int_0^1 y(s)ds. \ \square$$

The great majority of integral equations of the first kind encountered in applications have square integrable kernels and hence generate operators on L^2 which are compact. Solving such equations, in the generalized sense above, then involves the operator $K^{\dagger}$ and the solution process is stable if and only if $R(K)$ is closed. Now, it is easy to see that if K is compact, then $R(K)$ is closed if and only if it is finite dimensional (see Exercise 4.12) and hence the only compact operators K for which $K^{\dagger}$ is bounded are those with finite dimensional range. In the context of integral equations this says that the only Fredholm integral equations of the first kind giving rise to well-posed problems on L^2 are those whose kernels are degenerate.

Exercise 4.13: Suppose that K is compact and $R(K)$ is closed. Show that $R(K)$ contains no infinite complete orthonormal set. □

For compact linear operators K we can give a convenient representation for $K^{\dagger}$ in terms of the singular system $\{v_j, u_j; \mu_j\}$ discussed in the previous section. Indeed, if $y \in \mathcal{D}(K^{\dagger})$, then

$$y = y_1 + y_2, \ \ y_1 \in R(K) \ \text{and} \ y_2 \in R(K)^{\perp}.$$

Since $u_j \in R(K)$, we then have $< y, u_j > = < y_1, u_j >$ for all j and hence the vector

$$x = \sum_{j=1}^{\infty} \frac{< y_1, u_j >}{\mu_j} v_j = \sum_{j=1}^{\infty} \frac{< y, u_j >}{\mu_i} v_j$$

exists by Picard's criterion and satisfies $Kx = y_1$ and $x \in N(K)^{\perp}$. Thus x is a least squares solution lying in $N(K)^{\perp}$, that is,

$$K^{\dagger} y = \sum_{j=1}^{\infty} \frac{< y, u_j >}{\mu_j} v_j. \tag{4.22}$$

This representation of $K^{\dagger} y$ shows very clearly that $K^{\dagger}$ is unbounded if $R(K)$ is infinite dimensional. Indeed, a perturbation in y of the form ϵu_n gives a new right hand side of the form

$$y^{\epsilon} = y + \epsilon u_n$$

satisfying $||y^{\epsilon} - y|| = \epsilon$. Yet the generalized solutions satisfy

$$||K^{\dagger} y - K^{\dagger} y^{\epsilon}|| = \frac{\epsilon}{\mu_n} \rightarrow \infty \text{ as } n \rightarrow \infty.$$

4.4 Bibliographic Notes

The basic theory of linear spaces and linear operators can be found in many books; [RN] is particularly recommended. [K] is an excellent source on integral equations at a rather advanced level. For more on generalized inverses, see [BIG], [N1], [Grol].

5 Some Methodology for Inverse Problems

Calculations are plentiful while ideas are few
Tony Rothman, Science à la Mode

There is a huge and growing literature on methods for approximating solutions of inverse problems. In an introductory work it is not possible to do justice to the wide range of ideas and techniques currently in use. We have therefore chosen to illustrate only a few main themes in this chapter. We concentrate on regularization methods, the Backus-Gilbert approach to estimation, the maximum entropy idea, algebraic reconstruction techniques and output least squares.

Our aim in this chapter is to introduce the ideas behind the methods rather than to present the latest techniques for numerical solution of inverse problems. Our treatment is therefore rather theoretical and the hope is that it will provide the reader with the appropriate background for intelligent assessment of methods for solution of inverse problems.

5.1 The Method of Regularization

The idea of the method of regularization is to replace an ill-posed Fredholm integral equation of the first kind

$$\int_0^1 k(s,t)x(t)dt = y(s) \tag{5.1}$$

by a nearby well-posed Fredholm integral equation of the second kind. We will express equation (5.1) abstractly, as we did in the previous chapter, as an equation of the form

$$Kx = y \tag{5.2}$$

where K is a compact linear operator from a Hilbert space H_1 into a Hilbert space H_2. We have seen that generally equation (5.2) does not have a unique solution, therefore we seek a particular generalized solution, namely the least squares solution of minimum norm. That is, we assume that $y \in \mathcal{D}(K^\dagger)$ and our aim is to approximate $K^\dagger y$. We know that, ignoring the trivial case in which the kernel $k(\cdot,\cdot)$ is degenerate, the generalized solution $K^\dagger y$ depends discontinuously on y, but we would like our approximations to depend continuously on y. That is, our scheme involves exchanging the ill-posed problem for the exact solution for a well-posed problem for an approximate solution.

The generalized solution $x = K^{\dagger}y$ of (5.2) is a least squares solution and therefore it satisfies the normal equations

$$K^*Kx = K^*y \tag{5.3}$$

where K^* is the adjoint of K. Now, the self-adjoint compact operator K^*K has nonnegative eigenvalues and therefore, for any positive number α, the operator $K^*K + \alpha I$, where I is the identity operator on H_1, has strictly positive eigenvalues. In particular, the operator $K^*K + \alpha I$ has a bounded inverse, that is, the problem of solving the equation

$$(K^*K + \alpha I)x_\alpha = K^*y \tag{5.4}$$

is *well-posed*. The second kind equation (5.4) is called a *regularized* form of equation (5.3) and its unique solution

$$x_\alpha = (K^*K + \alpha I)^{-1}K^*y \tag{5.5}$$

is called the *Tikhonov* approximation to $K^{\dagger}y$, the minimum norm solution of (5.3).

The first order of business in studying these Tikhonov approximations is to show that they converge to $K^{\dagger}y$ as $\alpha \to 0$. This can be accomplished conveniently in terms of a singular system $\{v_j, u_j; \mu_j\}$ for K. Recall that $\{v_j\}$ is a complete orthonormal set for $N(K)^{\perp}$, $\{u_j\}$ is a complete orthonormal set for $\overline{R(K)}$, $\mu_j \to 0$, and

$$Kv_j = \mu_j u_j, \quad K^*u_j = \mu_j v_j. \tag{5.6}$$

From (5.4) we see that $\alpha x_\alpha = K^*y - K^*Kx_\alpha$ and hence

$$x_\alpha \in R(K^*) \subseteq N(K)^{\perp}.$$

Therefore, we may expand x_α in terms of the singular vectors $\{v_j\}$:

$$x_\alpha = \sum_{j=1}^{\infty} < x_\alpha, v_j > v_j.$$

Similarly,

$$\begin{aligned} K^*y &= \sum_{j=1}^{\infty} < K^*y, v_j > v_j = \sum_{j=1}^{\infty} < y, Kv_j > v_j \\ &= \sum_{j=1}^{\infty} \mu_j < y, u_j > v_j \end{aligned}$$

and substituting these results into (5.4) we find (using (5.6))

$$\sum_{j=1}^{\infty} (\mu_j^2 + \alpha) < x_\alpha, v_j > v_j = \sum_{j=1}^{\infty} \mu_j < y, u_j > v_j.$$

Therefore,

$$< x_\alpha, v_j >= \frac{\mu_j}{\mu_j^2 + \alpha} < y, u_j >$$

and hence

$$x_\alpha = \sum_{j=1}^{\infty} \frac{\mu_j}{\mu_j^2 + \alpha} < y, u_j > v_j.$$

The true minimum norm least squares solution is, according to equation (4.22),

$$K^\dagger y = \sum_{j=1}^{\infty} \frac{1}{\mu_j} < y, u_j > v_j.$$

Therefore,

$$\|x_\alpha - K^\dagger y\|^2 = \sum_{j=1}^{\infty} \left(\frac{\alpha}{\mu_j(\mu_j^2 + \alpha)} \right)^2 | < y, u_j > |^2. \tag{5.7}$$

Now, since

$$\left(\frac{\alpha}{\mu_j(\mu_j^2 + \alpha)} \right)^2 | < y, u_j > |^2 \leq \frac{1}{\mu_j^2} | < y, u_j > |^2$$

and

$$\sum_{j=1}^{\infty} \frac{1}{\mu_j^2} | < y, u_j > |^2 = \|K^\dagger y\|^2 < \infty,$$

we may, in passing to the limit as $\alpha \to 0$ in (5.7), interchange the limit and summation, giving

$$\lim_{\alpha \to 0} \|x_\alpha - K^\dagger y\|^2 = 0.$$

The vectors $\{x_\alpha\}$ are therefore bona fide approximations to $K^\dagger y$ in the sense that

$$x_\alpha \to K^\dagger y \text{ as } \alpha \to 0.$$

Moreover, since for each fixed $\alpha > 0$, the operator $(K^*K + \alpha I)^{-1} K^*$ is bounded, we see that the Tikhonov approximation x_α depends continuously on y, for each fixed $\alpha > 0$.

To summarize, in Tikhonov regularization, we approximate the minimum norm least squares solution $K^\dagger y$, which depends discontinuously on y, by a vector x_α, depending on a *regularization parameter* $\alpha > 0$, which is a continuous function of y. To put it another way, an ill-posed problem is approximated by a family of nearby well-posed problems.

Exercise 5.1: Suppose that $y = K^*Kw$, for some w. Use (5.7) to show that $\|x_\alpha - x\| = O(\alpha)$. □

Exercise 5.2: Let $H_1 = H_2 = L^2[0,1]$. Define $K : H_1 \to H_2$ by

$$(Kx)(s) = \int_0^s x(t)dt.$$

Suppose that y is absolutely continuous and $y(0) = 0$. Show that the Tikhonov approximation x_α for the problem $Kx = y$ is a solution of the boundary value problem

$$\alpha x''_\alpha(t) - x_\alpha(t) = -y'(t), \quad x_\alpha(1) = x'_\alpha(0) = 0.$$

Solve this for x_α and show that the solution depends continuously on y. □

In our models in Chapter 2 we saw that the function y in (5.1) is typically a measured or observed quantity and hence in practice the true y is not at our disposal. The best we can hope for is some estimate y^δ of y satisfying

$$\|y - y^\delta\| \leq \delta \tag{5.8}$$

where δ is a known bound on the measurement error. Instead of forming (5.5) with the true y, we must make do with the estimated data y^δ and form the regularized approximations

$$x_\alpha^\delta = (K^*K + \alpha I)^{-1}K^*y^\delta. \tag{5.9}$$

Now, we know that the approximations x_α using "clean" data y converge to the minimum norm least squares solution $K^\dagger y$ and therefore it is reasonable to compare x_α^δ to x_α:

$$x_\alpha^\delta - x_\alpha = (K^*K + \alpha I)^{-1}K^*(y^\delta - y).$$

From this we find (since $(K^*K + \alpha I)^{-1}K^* = K^*(KK^* + \alpha I)^{-1})$)

$$\begin{aligned}
\|x_\alpha^\delta - x_\alpha\|^2 &= < K^*(KK^* + \alpha I)^{-1}(y^\delta - y),\ K^*(KK^* + \alpha I)^{-1}(y^\delta - y) > \\
&= < (KK^* + \alpha I)^{-1}(y^\delta - y),\ KK^*(KK^* + \alpha I)^{-1}(y^\delta - y) > .
\end{aligned}$$

But, by the spectral mapping theorem,

$$\|KK^*(KK^* + \alpha I)^{-1}\| \leq 1 \text{ and } \|(K^*K + \alpha I)^{-1}\| \leq 1/\alpha,$$

and hence

$$\|x_\alpha^\delta - x_\alpha\| \leq \delta/\sqrt{\alpha}. \tag{5.10}$$

This inequality represents a stability bound for the approximation x_α^δ. It illustrates the classic dilemma in the analysis of ill-posed problems: for fixed $\delta > 0$, the bound blows up as $\alpha \to 0$, mirroring the instability in the underlying problem.

These considerations show that, for a fixed error level δ, letting the regularization parameter α approach zero generally results in an unstable process. Choosing a suitable regularization parameter, based on the error in the data, then becomes the crux of the matter. Following Tikhonov, we say that a choice $\alpha = \alpha(\delta)$ leads to a *regular* algorithm for the ill-posed problem (5.2) if

$$\alpha(\delta) \to 0 \text{ and } x^{\delta}_{\alpha(\delta)} \to K^{\dagger}y \text{ as } \delta \to 0.$$

Since

$$\begin{aligned} ||x^{\delta}_{\alpha(\delta)} - K^{\dagger}y|| &\leq ||x^{\delta}_{\alpha(\delta)} - x_{\alpha(\delta)}|| + ||x_{\alpha(\delta)} - K^{\dagger}y|| \\ &\leq \delta/\sqrt{\alpha(\delta)} + ||x_{\alpha(\delta)} - K^{\dagger}y|| \end{aligned}$$

and since we have shown that $x_{\alpha(\delta)} \to K^{\dagger}y$ as $\alpha(\delta) \to 0$, we see that the condition

$$\delta^2/\alpha(\delta) \to 0 \text{ as } \delta \to 0$$

is sufficient to ensure that Tikhonov's method gives a regular algorithm for (5.2).

Exercise 5.3: Show that if $y \in R(K^*K)$ and $\alpha = C\delta^{2/3}$ then $||x^{\delta}_{\alpha} - K^{\dagger}y|| = O(\delta^{2/3})$ for all y^{δ} satisfying (5.8). (Hint: See Exercise 5.1). □

Tikhonov's method also has a very important variational interpretation. Remember that the idea of the method is to approximate the generalized solution in a stable way. A reasonable way to attempt to do this is to minimize an augmented least squares functional

$$F_{\alpha}(x) = ||Kx - y^{\delta}||^2 + \alpha||x||^2. \tag{5.11}$$

In this functional the first term, when small, guarantees that x is "nearly" a least squares solution, while the second term tends to damp out wild instabilities in x. Now, the functional F_{α} in (5.11) actually achieves a minimum on H_1. The easiest way to see this is to note that if we define a norm $|\cdot|$ on the Hilbert space $H_1 \times H_2$ by

$$|\{u, v\}|^2 = ||v||^2 + \alpha||u||^2$$

(check that this actually is a norm on the product space $H_1 \times H_2$), then (5.11) measures the (squared) distance of the vector $\{0, y^{\delta}\} \in H_1 \times H_2$ from the graph of K, which is a closed convex set in $H_1 \times H_2$. Therefore, there is a vector $x \in H_1$ minimizing (5.11).

Any minimizer z of (5.11) must satisfy

$$\frac{d}{dt}\{||K(z + tw) - y^{\delta}||^2 + \alpha||z + tw||^2\}|_{t=0} = 0 \tag{5.12}$$

for all $w \in H_1$. Expressing the squared norms in terms of the inner product and expanding the quadratic forms we find that (5.12) is equivalent to

$$< Kz - y^{\delta}, Kw > +\alpha < z, w >= 0$$

or

$$< (K^*K + \alpha I)z - K^*y^{\delta}, w >= 0$$

for all $w \in H_1$. That is

$$(K^*K + \alpha I)z = K^*y^\delta.$$

We therefore see that the unique minimizer of the augmented least squares functional (5.11) is

$$x_\alpha^\delta = (K^*K + \alpha I)^{-1}K^*y^\delta \tag{5.13}$$

which is the Tikhonov approximation discussed previously. The variational characterization of the Tikhonov approximation (5.13) as the minimizer of the *Tikhonov functional* (5.11) has important analytical and computational implications for Tikhonov's method.

According to condition (5.10), any a priori choice $\alpha = \alpha(\delta)$ of the regularization parameter satisfying $\delta^2/\alpha(\delta) \to 0$ as $\delta \to 0$ leads to a regular algorithm for the solution of $Kx = y$. Although this asymptotic result may be theoretically satisfying, it would seem that a choice of the regularization parameter that is based on the actual computations performed, that is, an *a posteriori* choice of the regularization parameter would be more effective in practice. One such a posteriori strategy is the discrepancy principle of Morozov. The idea of the strategy is to choose the regularization parameter so that the size of the residual $||Kx_\alpha^\delta - y^\delta||$ is the same as the error level in the data:

$$||Kx_\alpha^\delta - y^\delta|| = \delta. \tag{5.14}$$

The following exercise gives some insight into the choice of the regularization parameter by the discrepancy principle.

Exercise 5.4: Show that if $||x||$ is a minimum subject to the constraint $||Kx - y^\delta|| \leq \delta$, then $||Kx - y^\delta|| = \delta$. □

Assuming that the signal-to-noise ratio is larger than one, that is, $||y^\delta|| > \delta$, and that $y \in R(K)$, then it is not hard to see that there is a unique positive parameter α satisfying (5.14). To do this, we use the singular value decomposition:

$$||Kx_\alpha^\delta - y^\delta||^2 = \sum_{j=1}^{\infty} \left(\frac{\alpha}{\mu_j^2 + \alpha}\right)^2 |< y^\delta, u_j >|^2 + ||Py^\delta||^2 \tag{5.15}$$

where P is the orthogonal projector of H_2 onto $R(K)^\perp$. From (5.15) we see that the real function

$$f(\alpha) = ||Kx_\alpha^\delta - y^\delta||$$

is a continuous, increasing function of α satisfying (since $Py = 0$)

$$\lim_{\alpha \to 0+} f(\alpha) = ||Py^\delta|| = ||Py^\delta - Py|| \leq ||y^\delta - y|| \leq \delta$$

and

$$\lim_{\alpha \to \infty} f(\alpha) = \|y^\delta\| > \delta.$$

Therefore, by the intermediate value theorem, there is a unique $\alpha = \alpha(\delta)$ satisfying (5.14). This choice of the regularization parameter is called the choice by the *discrepancy method.*

We close this section by showing that the choice $\alpha(\delta)$ as given by the discrepancy method (5.14) leads to a regular scheme for approximating $K^\dagger y$, that is

$$x^\delta_{\alpha(\delta)} \to K^\dagger y \text{ as } \delta \to 0.$$

To do this it is sufficient to show that for *any* sequence $\delta_n \to 0$ there is a subsequence, which for notational convenience we will denote by $\{\delta_k\}$, such that $x^{\delta_k}_{\alpha(\delta_k)} \to K^\dagger y$. We are assuming that $y \in R(K)$ and to simplify notation we set $x = K^\dagger y$. Then x is the unique vector satisfying $Kx = y$ and $x \in N(K)^\perp$.

From the variational characterization of the Tikhonov approximation we have

$$F_{\alpha(\delta)}(x^\delta_{\alpha(\delta)}) \leq F_{\alpha(\delta)}(x)$$

that is,

$$\begin{aligned} \delta^2 + \alpha(\delta)\|x^\delta_{\alpha(\delta)}\|^2 &= \|Kx^\delta_{\alpha(\delta)} - y^\delta\|^2 + \alpha(\delta)\|x^\delta_{\alpha(\delta)}\|^2 \\ &\leq F_{\alpha(\delta)}(x) = \|y - y^\delta\|^2 + \alpha(\delta)\|x\|^2 \\ &\leq \delta^2 + \alpha(\delta)\|x\|^2 \end{aligned}$$

and hence $\|x^\delta_{\alpha(\delta)}\| \leq \|x\|$. Therefore, for any sequence $\delta_n \to 0$ there is a subsequence $\delta_k \to 0$ with $x^{\delta_k}_{\alpha(\delta_k)} \rightharpoonup z$, for some z. Since

$$x^\delta_{\alpha(\delta)} = K^*(KK^* + \alpha(\delta)I)^{-1}y^\delta \in R(K^*) \subseteq N(K)^\perp$$

and $N(K)^\perp$ is weakly closed, we find that $z \in N(K)^\perp$. Also, since

$$\|Kx^{\delta_k}_{\alpha(\delta_k)} - y^{\delta_k}\| \to 0$$

we see that $Kx^{\delta_k}_{\alpha(\delta_k)} \to y$. But K is weakly continuous and therefore $Kx^{\delta_k}_{\alpha(\delta_k)} \rightharpoonup Kz$. It follows that $Kz = y$ and $z \in N(K)^\perp$, i.e., $z = x$. Since $\|x^{\delta_k}_{\alpha(\delta_k)}\| \leq \|x\|$, we then have

$$\|x\|^2 = \lim_{k \to \infty} < x^{\delta_k}_{\alpha(\delta_k)}, x > \leq \underline{\lim}_{k \to \infty} \|x^{\delta_k}_{\alpha(\delta_k)}\| \cdot \|x\|$$

and therefore $x^{\delta_k}_{\alpha(\delta_k)} \rightharpoonup x$ and $\|x^{\delta_k}_{\alpha(\delta_k)}\| \to \|x\|$ and hence $x^{\delta_k}_{\alpha(\delta_k)} \to x$, and the proof is complete.

5.2 Discretization Methods

The computational solution of an integral equation of the first kind

$$\int_a^b k(s,t)x(t)dt = y(s) \tag{5.16}$$

requires that the problem be *discretized*, that is, expressed in terms of finitely many unknowns. The simplest way to accomplish this is to apply some quadrature rule, like the midpoint rule, Simpson's rule, etc., to the integral. Applying a quadrature rule with weights $\{w_j\}_{j=1}^n$ and notes $\{t_j\}_{j=1}^n$ to (5.16) we obtain the approximate problem

$$\sum_{j=1}^n w_j k(s,t_j)x_j = y(s) \tag{5.17}$$

where the numbers x_j are approximations to $x(t_j)$. Now (5.17) still represents an infinite system in that a constraint is specified for each of infinitely many values of s. Of course we can convert (5.17) into a finite dimensional problem by *collocation*, that is, by requiring (5.17) to hold at certain specified collocation points $\{s_i\}_{i=1}^m$:

$$\sum_{j=1}^n w_j k(s_i,t_j)x_j = y(s_i), \quad i = 1,...,m. \tag{5.18}$$

In this way the integral equation (5.16) is approximated by the $m \times n$ linear system

$$Ax = b \tag{5.19}$$

where A is the $m \times n$ matrix with entries $[w_j k(s_i,t_j)]$, x is now an n-vector which is meant to approximate

$$[x(t_1),...,x(t_n)]^T$$

and

$$b = [y(s_1),...,y(s_m)]^T.$$

In discretizing an ill-posed integral equation of the first kind an ill-conditioned linear system is produced. Generally, the finer the discretization, the closer the algebraic problem approximates the ill-posed continuous problem and hence the more ill-conditioned the algebraic problem becomes. We illustrate this with an example of Fox and Goodwin (see [B, p.665]):

$$\int_0^1 (s^2+t^2)^{1/2} f(t)dt = \frac{1}{3}[(1+s^2)^{3/2} - s^3]. \tag{5.20}$$

This problem has the exact solution $f(t) = t$. Suppose we produce a simple discretization of (5.16) by applying the midpoint rule with gridsize $h = 1/n$ to the integral and collocating at the midpoints, that is,

$$\frac{1}{n}\sum_{j=1}^{n}(t_i^2 + t_j^2)^{1/2}x_j = b_i, \quad i = 1, ..., n$$

where $t_j = (2j-1)/(2n)$, $b_i = \frac{1}{3}[(1+t_i^2)^{3/2} - t_i^3]$, and $x_j \approx f(t_j)$, $i, j = 1, .., n$. We are then led to an $n \times n$ algebraic system

$$Ax = b \tag{5.21}$$

where $a_{ij} = \frac{1}{n}(t_i^2 + t_j^2)^{1/2}$. As n increases, we expect that the matrix A will more closely represent the kernel in (5.16) and hence will become more ill-conditioned. In fact, MATLAB computations of the condition number of the matrix A yield the results in Figure 5.1, confirming our suspicions.

n	cond(A)
2	9.9
10	1.9×10^{10}
20	3.2×10^{17}
50	3.3×10^{19}

Figure 5.1

With such high condition numbers, we should expect that the computed solution of (5.21) will approximate the true solution $f(t) = t$ of (5.20) poorly, even though the error in b is attributable only to machine rounding. In fact, using $n = 10$ in the discretization we plot in Figure 5.2 the true solution of (5.20) (solid) along with the (interpolated) computed solution (dashed) of (5.21).

With a well-posed problem, we are accustomed to getting better results as we refine the discretization. However, for an integral equation of the first kind, refining the discretization causes the discrete problem to more closely mirror the ill-posed nature of the continuous problem. For example, repeating the numerical experiment of solving (5.21) with $n = 20$ gives the results in Figure 5.3. We see that the approximate solution is now quite "off the scale" and is of no use whatsoever in approximating the true solution $f(t) = t$.

In trading the integral equation (5.20) for the linear system (5.21) we are still faced with a problem that may have no solution (if $b \notin R(A)$), may have more than one solution (if $N(A) \neq \{0\}$) and which is singular or ill-conditioned, with the degree of ill-conditioning increasing as the dimensions m and n increase. In solving (5.21) one then encounters essentially the same issues of existence, uniqueness and stability

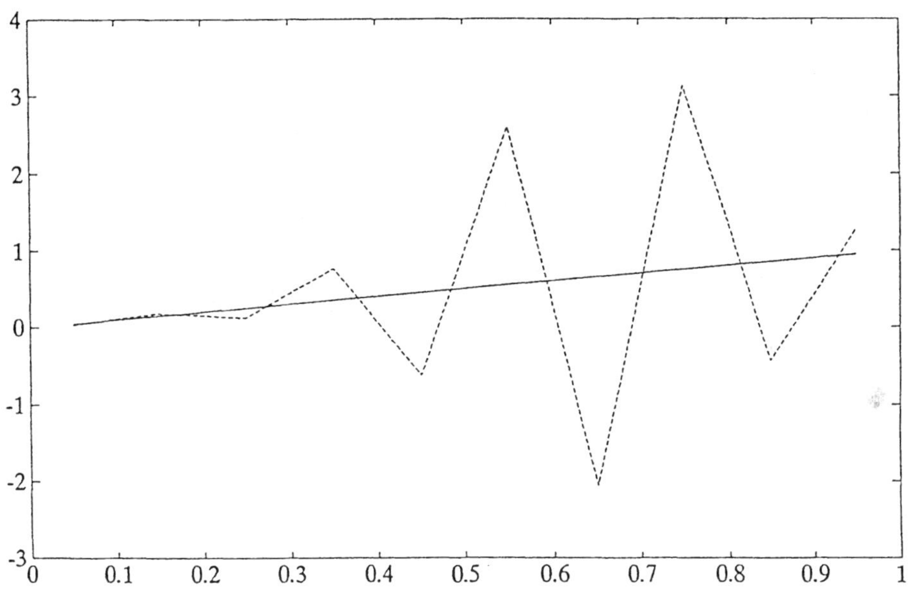

Figure 5.2

discussed previously for infinite dimensional problems. Some type of regularization, for example, Tikhonov regularization, must then be applied to the discrete problem if we hope to get reasonable results. It is surprising how much a modest amount of regularization can help. To illustrate this we again consider the discretization (5.21) of (5.20) using $n = 20$, but now we add (uniform) random errors ϵ_i to the right hand side with $|\epsilon_i| \leq \delta = 10^{-4}$. In Figure 5.4 are plotted the true solution (solid), the solution obtained using Tikhonov regularization with $\alpha = 2.5 \times 10^{-9}$ (*) and the solution obtained using Tikhonov regularization with regularization parameter chosen by the discrepancy method (dashed).

The variational characterization of the Tikhonov approximations suggest another way to discretize the problem. Recall that x_α^δ has been characterized as the unique vector which minimizes the Tikhonov functional

$$F_\alpha(u) = ||Ku - y^\delta||^2 + \alpha||u||^2 \tag{5.22}$$

over the Hilbert space H_1. To turn this into a finite dimensional problem, we might simply try to minimize (5.22) over a finite dimensional subspace of H_1. For example, if V_n is an n-dimensional subspace of H_1 spanned by the linearly independent vectors $\{v_1, ..., v_n\}$, then the vector $u^{(n)} \in V_n$ minimizes F_α over V_n if and only if

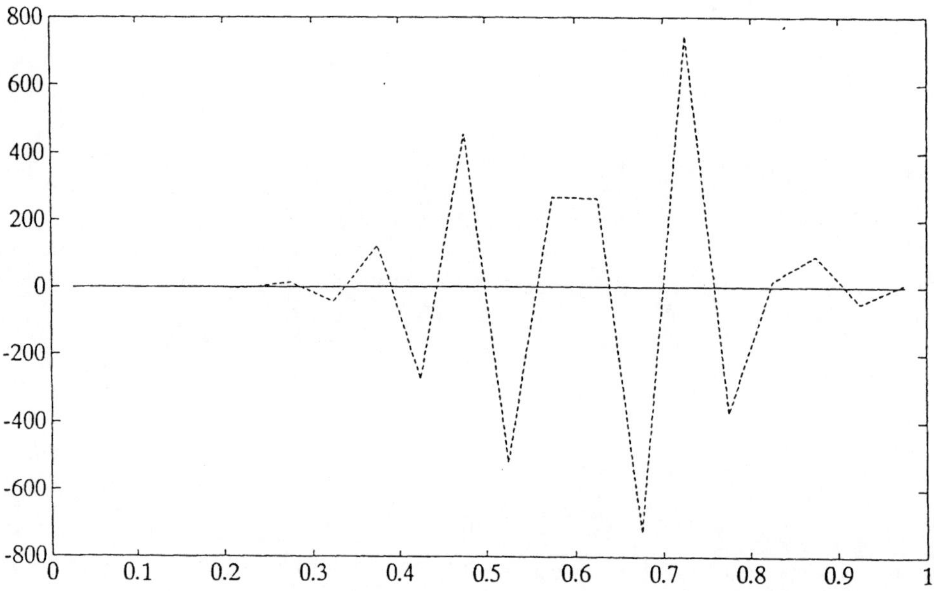

Figure 5.3

$$u^{(n)} = \sum_{j=1}^{n} x_j v_j \tag{5.23}$$

where the coefficients $\{x_j\}$ give a stationary point of the functional F_α.

Exercise 5.4: Show that the vector $u^{(n)}$ in (5.23) minimizes F_α over V_n if and only if the vector

$$x = [x_1, ..., x_n]^T \in \mathbf{R}^n$$

satisfies the equation

$$(A + \alpha B)x = w \tag{5.24}$$

where

$$w = [< Kv_1, y^\delta >, ..., < Kv_n, y^\delta >]^T,$$

the $n \times n$ matrix A has entries

$$a_{ij} =< Kv_i, Kv_j >, \quad i, j = 1, ..., n,$$

and the matrix B has entries

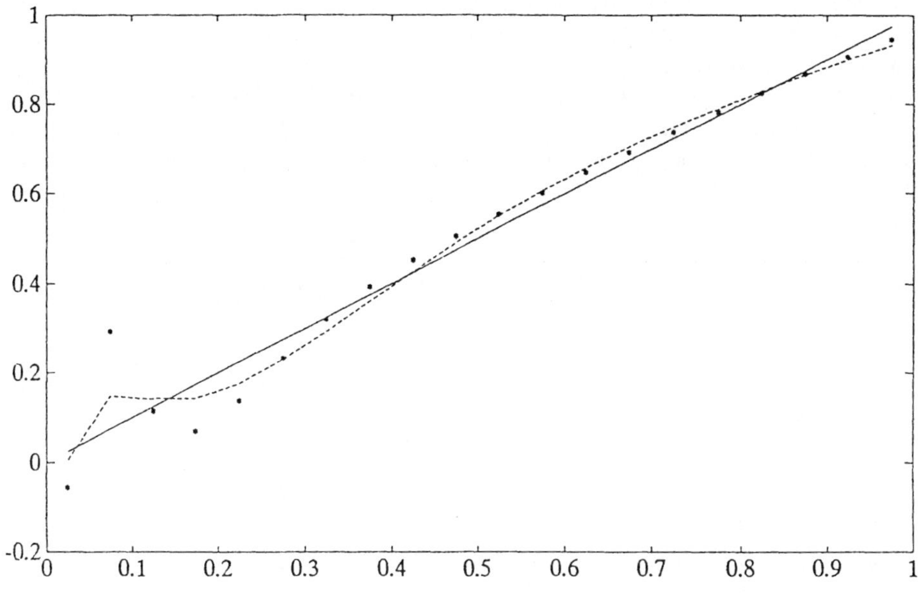

Figure 5.4

$$b_{ij} = < v_i, v_j >, \quad i, j = 1, ..., n. \ \Box$$

We might call this type of discretization of a "finite element" discretization because the computed numbers $\{x_j\}$ are coefficients of certain basis functions $\{v_j\}$, which often will be taken as basic spline functions on some grid. For example, we might take the $\{v_j\}$ to be the piecewise linear "tent" functions pictured in Figure 3.13. Two remarks on this method are in order. First, in this method the discretization and regularization work hand-in-hand to produce the system (5.24), unlike in the previous method in which the discretization, by quadrature and collocation, takes place first and then the regularization is performed after the discretization is complete. Second, it is clear that the matrices in (5.24) are much more expensive to compute than the matrix A in (5.21). Indeed, the matrix A in (5.21) is computed by simply evaluating the kernel at certain points (and multiplying by appropriate weights). On the other hand, the matrices A and B (5.24) are generated by computing the inner products

$$< Kv_i, Kv_j > \quad \text{and} \quad < v_i, v_j > .$$

This requires the more expensive operations of applying the operator K to certain functions and the computation of integrals (the inner-products). In practice these integrals would be computed by some high order quadrature rules, for example, Gaussian quadrature.

We close this section by suggesting a third way in which Tikhonov regularization can be turned into a finite-dimensional problem. In this method the regularization is performed first. The regularized approximation x_α^δ, as discussed in the previous section, satisfies

$$(K^*K + \alpha I)x_\alpha^\delta = K^*y^\delta. \tag{5.25}$$

Now, as we saw in Exercise 4.3, K^*K is itself an integral operator,

$$(K^*Kx)(s) = \int_0^1 \tilde{k}(s,t)x(t)dt$$

where the kernel $\tilde{k}(s,t)$ is given by

$$\tilde{k}(s,t) = \int_0^1 k(u,s)k(u,t)du.$$

Suppose that we apply a quadrature rule to the integral defining the kernel $\tilde{k}(\cdot,\cdot)$:

$$\int_0^1 k(u,s)k(u,t)du \approx \sum_{j=1}^n w_j k(u_j,s)k(u_j,t).$$

But the kernel

$$\sum_{j=1}^n w_j k(u_j,s)k(u_j,t) \tag{5.26}$$

is a finite sum of products of functions of s and t alone, i.e. it is a degenerate kernel. Replacing K^*K in (5.25) with the finite rank operator generated by the degenerate kernel (5.26) then results in a Fredholm integral equation of the second kind with degenerate kernel. Such an equation is equivalent to a finite dimensional linear system and hence may be solved by algebraic means (see [B]).

5.3 Iterative Methods

Iterative methods for solving equations are popular because they require only relatively simple operations to be performed repeatedly. There are many iterative methods that can be, and are, applied to ill-posed problems. In this section we treat only the simplest of these methods, *Landweber-Fridman* iteration.

Suppose that K is a compact linear operator and $y \in \mathcal{D}(K^\dagger)$. Recall that the generalized solution $x = K^\dagger y$ that we seek is the unique vector in $N(K)^\perp$ which satisfies the equation

$$K^*Kx = K^*y. \tag{5.27}$$

We take (5.27) as our starting point, multiply by a positive factor β (the role of β will become apparent later) and rewrite (5.27) as

$$x = x + \beta(K^*y - K^*Kx).$$

This suggests the iterative method

$$x_{n+1} = x_n + \beta(K^*y - K^*Kx_n). \tag{5.28}$$

Note that if we take $x_0 = 0$, then by (5.28) we have $x_n \in R(K^*) \subseteq N(K)^\perp$ for every n. Therefore, if $\{x_n\}$ converges to a vector x, then $x \in N(K)^\perp$ and x satisfies (5.27), that is, $x = K^\dagger y$. Therefore, it remains to find a condition on β that will guarantee the convergence of $\{x_n\}$. Let $x = K^\dagger y$ and denote the error in the approximation x_n by e_n:

$$e_n = x_n - x.$$

From (5.28) and (5.27) we obtain

$$e_{n+1} = (I - \beta K^*K)e_n,$$

where I is the identity operator. Therefore,

$$e_n = (I - \beta K^*K)^n e_0. \tag{5.29}$$

Now let $\|K\|^2 = \lambda_1 \geq \lambda_2 \geq \ldots$ be the nonzero eigenvalues of K^*K and let $\{v_j\}$ be a corresponding system of orthonormal eigenvectors. Then $\{v_j\}$ is a complete orthonormal system for $N(K^*K)^\perp = N(K)^\perp$, and, since $e_0 = -x \in N(K)^\perp$, we may expand e_0 in terms of the eigenvectors $\{v_j\}$. We then have from (5.29)

$$\|e_n\|^2 = \sum_{j=1}^{\infty}(1 - \beta\lambda_j)^{2n}| < e_0, v_j > |^2. \tag{5.30}$$

Suppose that β is chosen so that $0 < \beta < 2/\lambda_1$. Then

$$|1 - \beta\lambda_j| < 1$$

for all j and, by Bessel's inequality,

$$\sum_{j=1}^{\infty} | < e_0, v_j > |^2 \leq \|e_0\|^2.$$

Since $|1 - \beta\lambda_j|^{2n} \to 0$ as $n \to \infty$, for each j, we see from (5.30) that $\|e_n\|^2 \to 0$, that is, $x_n \to K^\dagger y$.

Exercise 5.5: Show that for arbitrary x_0, the Landweber-Fridman method (5.28) converges to the least squares solution of $Kx = y$ which is nearest to x_0. □

Since the eigenvalues $\{\lambda_j\}$ of K^*K converge to zero, we see from (5.30) that for larger values of j the damping factors $(1-\beta\lambda_j)^{2n}$ are close to 1. Therefore, in the iteration process the components of the solution corresponding to low order eigenvectors are resolved first, while the higher order eigenvectors take much longer to make their effect known. In fact, the slow rate of convergence of the Landweber-Fridman method is its major drawback.

Exercise 5.6: Suppose $x_0 = 0$ and $K^\dagger y \in R(K^*K)$. Use (5.30) to show that $||x_n - K^\dagger y|| = O(\frac{1}{n})$. □

Consider now the influence of error in the data. Suppose that the available data is a vector y^δ satisfying

$$||y - y^\delta|| \leq \delta.$$

Using the vector y^δ in the iterative method we generate the approximations $x_0 = 0$,

$$x_{n+1}^\delta = x_n^\delta + \beta(K^*y^\delta - K^*Kx_n^\delta). \tag{5.31}$$

Our aim is to show that, in this iterative method, it is the parameter n that plays the role of the regularization parameter. That is, there is a choice of the "stopping value" $n = n(\delta)$, with the property that, if the iteration is terminated at step $n(\delta)$, then

$$x_{n(\delta)}^\delta \to K^\dagger y \text{ as } \delta \to 0.$$

As with Tikhonov regularization, we establish this by estimating the stability error

$$d_n^\delta = x_n^\delta - x_n.$$

From (5.28) and (5.31), we find

$$d_{n+1}^\delta = (I - \beta K^*K)d_n^\delta + \beta K^*(y^\delta - y), \quad d_0^\delta = 0.$$

Since, by choice of β, $||I - \beta K^*K|| \leq 1$, we have

$$||d_{n+1}^\delta|| \leq ||d_n^\delta|| + \beta||K||\delta,$$

and hence

$$||d_n^\delta|| \leq n\beta||K||\delta.$$

We therefore have

$$\begin{aligned} ||x_n^\delta - K^\dagger y|| &\leq ||x_n - K^\dagger y|| + ||x_n^\delta - x_n|| \\ &\leq ||x_n - K^\dagger y|| + O(n\delta). \end{aligned}$$

But we have already shown that $x_n \longrightarrow K^\dagger y$ and hence a sufficient condition for regularity of $x^\delta_{n(\delta)}$ is that the iteration number $n = n(\delta)$ satisfy $n\delta \longrightarrow 0$ as $\delta \longrightarrow 0$.

Exercise 5.7: Show that if $x_0 = 0$, $K^\dagger y \in R(K^*K)$ and $n = [\delta^{-1/2}]$, then $\|x_n^\delta - K^\dagger y\| = O(\sqrt{\delta})$. □

In general, we expect that the Landweber-Fridman iteration method converges slowly, and if there is error in the data, then continued iteration past an appropriate stopping point only causes a magnification of the error components associated with small singular values. This is illustrated for the Fox-Goodwin example (5.20) in Figure 5.5. In the computations a twenty point mid-point rule discretization was taken and the right hand side was contaminated with 10% uniform random error. The figure shows the straight line true solution, the approximation solution computed using 100 iterations of Landweber's method (solid) and the solution computed using 10,000 iterations of Landweber's method ($*$).

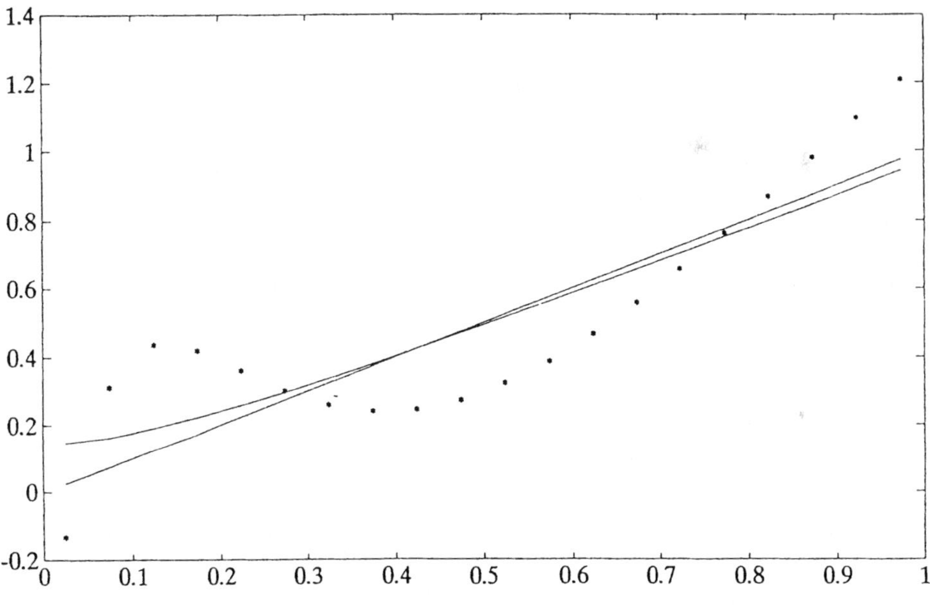

Figure 5.5

It is easy to see from (5.28) that

$$x_{n+1} = P_n(K^*K)K^*y \tag{5.32}$$

where $P_n(\lambda)$ is a certain polynomial of degree n. This raises the question of choosing a polynomial P_n of degree n so as to optimize the iteration in some sense. One approach would be to construct a polynomial so that the residual

$$||Kx_{n+1} - y||^2 \tag{5.33}$$

is as small as possible. Note that if $y \in R(K)$, then

$$\begin{aligned} ||Kx_{n+1} - y||^2 &= ||KP_n(K^*K)K^*Kx - Kx||^2 \\ &= < K^*K(I - K^*KP_n(K^*K))^2x, x > \end{aligned} \tag{5.34}$$

where $x = K^\dagger y$. Therefore minimization of the residual (5.33) is equivalent to minimization of the expression (5.35). There is, in fact, an iterative method of the form (5.32) that minimizes (5.35) (see [Lu, p.247]). The method is called the *conjugate gradient* method. We will not discuss the conjugate gradient method and other methods of optimizing an iteration of the form (5.32), but some references for further reading are cited in the final section.

5.4 TSVD

A straightforward approach to computing $K^\dagger y$ is to truncate the singular value decomposition (4.22):

$$K^\dagger y = \sum_{j=1}^{\infty} \frac{< y, u_j >}{\mu_j} v_j . \tag{5.35}$$

We recall that the vectors $\{u_j\}$ form a complete orthonormal set for $\overline{R(K)}$, $\{v_j\}$ is a complete orthonormal set for $N(K)^\perp$, $\mu_j \to 0$ and

$$Kv_j = \mu_j u_j \text{ and } K^*u_j = \mu_j v_j .$$

That is, μ_j^2 are the nonzero eigenvalues of K^*K with associated orthonormal eigenvectors $\{v_j\}$ and μ_j^2 are also the nonzero eigenvalues of KK^* with associated orthonormal eigenvectors $\{u_j\}$.

If the expansion (5.35) is truncated at the level n to form the *truncated singular value decomposition* (TSVD)

$$x_n = \sum_{j=1}^{n} \frac{< y, u_j >}{\mu_j} v_j , \tag{5.36}$$

then by (5.35), $x_n \to K^\dagger y$ as $n \to \infty$.

Exercise 5.8: Show that if $K^\dagger y \in R(K^*K)$, then $||x_n - K^\dagger y|| = O(\mu_{n+1})$. □

Consider now the effect of error in the data. If the available data is y^δ where $||y - y^\delta|| \leq \delta$ and the TSVD approximation using the available data is

$$x_n^\delta = \sum_{j=1}^{n} \frac{< y^\delta, u_j >}{\mu_j} v_j, \tag{5.37}$$

then

$$x_n - x_n^\delta = \sum_{j=1}^{n} \frac{< y - y^\delta, u_j >}{\mu_j} v_j. \tag{5.38}$$

We can then estimate the stability error as follows:

$$\begin{aligned} ||x_n - x_n^\delta||^2 &= < \sum_{j=1}^{n} \mu_j^{-1} < y - y^\delta, u_j > v_j, \sum_{j=1}^{n} \mu_j^{-1} < y - y^\delta, u_j > v_j > \\ &= \sum_{j=1}^{n} \mu_j^{-2} | < y - y^\delta, u_j > |^2 \\ &\leq \mu_n^{-2} \sum_{j=1}^{n} | < y - y^\delta, u_j > |^2 \leq \delta^2 \mu_n^{-2}. \end{aligned}$$

Therefore,

$$||x_n^\delta - K^\dagger y|| \leq ||x_n - K^\dagger y|| + \delta \mu_n^{-1} \tag{5.39}$$

and hence if $n = n(\delta)$ is chosen so that $\delta \mu_n^{-1} \to 0$ as $\delta \to 0$, then $x_n^\delta \to K^\dagger y$. That is, for appropriate choice of the truncation level, the TSVD method is a regularization method.

Exercise 5.9: Suppose that $K^\dagger y \in R(K^*K)$ and that $||y - y^\delta|| \leq \delta$. Show that if $n = n(\delta)$ is chosen such that $\mu_{n+1}^2 \leq \delta < \mu_n^2$, then $||x_n^\delta - K^\dagger y|| = O(\sqrt{\delta})$. □

As with any regulariztion method for ill-posed problems, the choice of the truncation level in the TSVD method is a delicate matter. For a fixed level of error in the data, however small, the TSVD approximations will begin to diverge if the truncation level is increased. The reason for this is again the fact that the higher the truncation level, the nearer the finite dimensional problem approaches the ill-posed continuous problem. This is illustrated in Figure 5.6.

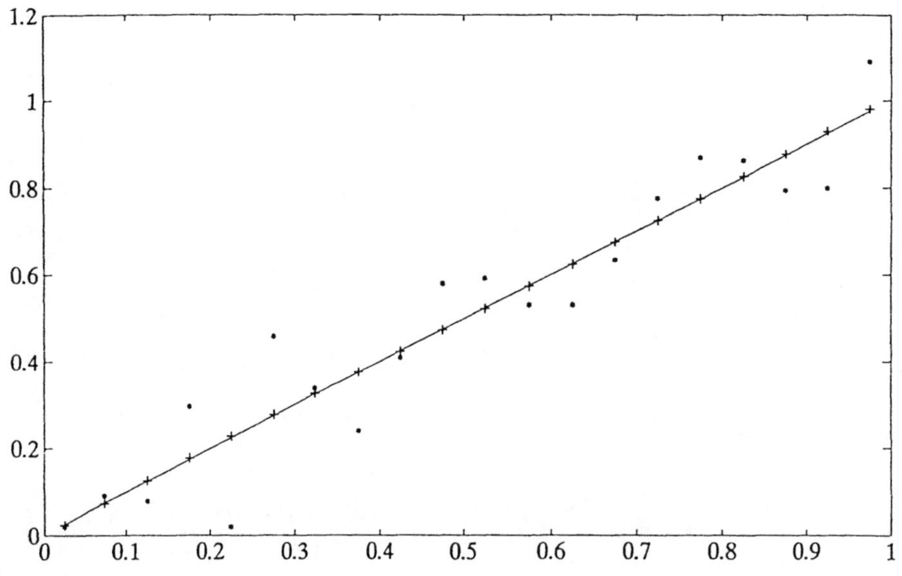

Figure 5.6

In this figure the results of applying the TSVD method to the twenty point midpoint rule discretization of the Fox-Goodwin example (5.20) are displayed. The only error in the right hand side is that due to machine rounding. The true solution $x(t) = t$ is displayed as the solid line, the TSVD solution with $n = 5$ is displayed as '+' and the TSVD solution with $n = 15$ is displayed as '$*$'.

5.5 The Maximum Entropy Method

The origins of the maximum entropy method for estimating solutions of inverse problems can be traced to the fundamental work of Boltzmann on statistical mechanics (the term *entropie* was coined by Clausius in 1865). Boltzmann analyzed a large number N of gas molecules by subdividing phase space into s congruent cells. The statistical state of such a system was then given by a partitioning $(N_1, N_2, ..., N_s)$ where

$$N_1 + N_2 + ... + N_s = N \tag{5.40}$$

and N_k is the number of molecules in the kth cell. The number of such distributions given a specified state is then

$$W = \frac{N!}{N_1!N_2!...N_s!},$$

and the state with the greatest number of distributions is found by maximizing W, subject to the constraint (5.40). But note that W is a maximum if and only if $N^{-1} \ln W$ is a maximum. Also,

$$\ln W = \ln N! - \sum_{k=1}^{s} \ln N_k!$$

and by Sterling's approximation for $n!$,

$$\lim_{n \to \infty} \frac{n!}{n(\ln n - 1)} = 1.$$

Therefore, assuming $N_1, N_2, ..., N_s$ are all large (remember, we are discussing gas molecules), we have

$$\begin{aligned} \ln W &\approx N(\ln N - 1) - \sum_{k=1}^{s} N_k(\ln N_k - 1) \\ &= N \ln N - \sum_{k=1}^{s} N_k \ln N_k \\ &= -\sum_{k=1}^{s} N_k \ln(N_k/N). \end{aligned}$$

It follows that

$$N^{-1} \ln W \approx -\sum_{k=1}^{s} p_k \ln p_k,$$

where $p_k = N_k/N$ represents the probability that a molecule occupies the kth cell in phase space. It is this value

$$H = -\sum_{k=1}^{s} p_k \ln p_k \tag{5.41}$$

that is called the *entropy* of the probability distribution $(p_1, p_2, ..., p_s)$. The distribution having maximum entropy then corresponds to the distribution of gas molecules having the maximum number of realizable microstates satisfying the constraint (5.40).

In general, the entropy function of a probability distribution measures the degree of uncertainty involved in trying to guess the exact state of a system having this distribution. Consider, for example, the simplest case in which there are two possible states with probabilities p and $1 - p$, respectively. The entropy is given by

$$H = -p \ln p - (1 - p) \ln(1 - p).$$

If $p = 0$ or $p = 1$, then there is complete certainty and the entropy is 0 (we take $0 \ln 0 = 0$). On the other hand, if $p = .5$, then the probability distribution is *uniform* and the uncertainty is at a maximum in the sense that there is no reason to choose either state over the other. In this case H takes its maximum value of $\ln 2$.

Exercise 5.10: Show that the finite discrete probability distribution $(p_1, p_2, ..., p_n)$ with maximum entropy is the uniform density $p_k = 1/n$, for $k = 1, 2, ..., n$. □

Exercise 5.11: Show that the infinite discrete probability density $(p_1, p_2, ...)$ with mean $\mu = \sum_{k=1}^{\infty} k p_k$ having maximum entropy is given by $p_k = \frac{1}{\mu+1}(\frac{\mu}{\mu+1})^k$. □

Our justification for using Boltzmann's entropy function as a measure of uncertainty was based on a use of Sterling's approximation. It turns out that this entropy measure is essentially unique in the sense that any uncertainty function satisfying certain "natural" axioms is essentially of the form (5.41). This axiomatic development of entropy was given by Claude Shannon in his monumental paper on information theory. Shannon assumed that any natural uncertainty function $H(p_1, ..., p_n)$ defined on a probability distribution should satisfy three conditions:

1. H is continuous
2. $A(n) = H(\frac{1}{n}, ..., \frac{1}{n})$ is an increasing function of n
3. H satisfies a natural partitioning property.

The property (2) simply indicates that for uniform distributions the uncertainty of choosing a given state increases as the number of possible states increases.

The partitioning property (3) refers to refinements of the slate space. For example, if there are two states ($n = 2$) with probabilities q_1 and q_2, then the uncertainty is some value $H(q_1, q_2)$. Suppose that the second state is refined into two substates with probabilities p_1 and p_2 as in Figure 5.7. The refined system then has three states with associated probabilities q_1, q_2p_1 and q_2p_2. The uncertainty of the refined system is assumed to satisfy

$$H(q_1, q_2p_1, q_2p_2) = H(q_1, q_2) + q_2 H(p_1, p_2).$$

This axiom conveys the meaning that, as the number of possible states is increased by refining the states, the uncertainty in choosing a given state increases by a probability weighted average of the uncertainties of the subsystems. In general, if in an n state system with probabilities $q_1, ..., q_n$, the kth state is refined to m_k states with probabilities $p_1^{(k)}, ..., p_{m_k}^{(k)}$, then the resulting partitioned system with $m_1 + ... + m_k$ states has uncertainty function given by

$$\begin{aligned} &H(q_1 p_1^{(1)}, ..., q_1 p_{m_1}^{(1)}, q_2 p_1^{(2)}, ..., q_2 p_{m_2}^{(2)}, ..., q_n p_1^{(n)}, ..., q_n p_{m_n}^{(n)}) \\ &= H(q_1, ..., q_n) + q_1 H(p_1^{(1)}, ..., p_{m_1}^{(1)}) + ... + q_n H(p_1^{(n)}, ..., p_{m_n}^{(n)}). \end{aligned} \quad (5.42)$$

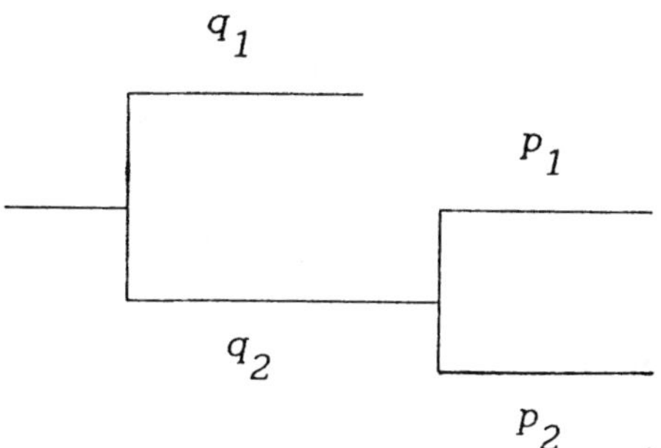

Figure 5.7

Consider now the function $H(q_1, ..., q_n)$ where each of the q_k is a rational number. We may then write

$$q_k = n_k/N, \quad N = \sum_{k=1}^{n} n_k$$

for some positive integers $\{n_k\}$. From (5.42) we then find

$$H(q_1, ..., q_n) + \sum_{j=1}^{n} q_j H(\frac{1}{n_j}, ..., \frac{1}{n_j}) = H(\frac{1}{N}, ..., \frac{1}{N})$$

or

$$H(q_1, ..., q_n) = A(N) - \sum_{j=1}^{n} q_j A(n_j). \tag{5.43}$$

Therefore, H is determined on rational values by the function A. Also, from (5.43), we find that if $n_j = m$ for all j, then

$$A(n) = A(mn) - A(m),$$

that is, the continuous function A satisfies one of Cauchy's famous functional equations:

$$A(mn) = A(m) + A(n).$$

Therefore, $A(x) = k \ln x$, for some constant k, which is positive by (2). From (5.43) we see that

$$\begin{aligned} H(q_1, ..., q_n) &= k(\ln N - \sum_{j=1}^{n} q_j \ln n_j) \\ &= k \sum_{j=1}^{n} q_j (N - \ln n_j) \\ &= -k \sum_{j=1}^{n} q_j \ln q_j \end{aligned}$$

for rational values q_j. Since H is continuous, it is determined by its values on the rationals and hence we find that any "uncertainty" functions satisfying (1), (2), (3) is a positive multiple of the entropy function.

The discussion above shows that the entropy function (5.41) provides a meaningful measure of the disorder of a system, or equivalently, the uncertainty involved in choosing a given state for a system. The *maximum entropy method* for inverse problems exploits this idea by invoking a kind of *principle of parsimony* in trying to reconstruct a solution of the problem. Namely, if the solution is known to be nonnegative, and hence may be normalized so that it is essentially a probability distribution, one chooses the distribution satisfying the given constraints which is maximally uncommitted with respect to missing information. To put it another way, one chooses the distribution which satisfies the given constraints and has maximum entropy. In the next exercise, the reader is asked to work out and compare a minimum norm reconstruction of an inverse problem in which the only known information is the mean of the probability distribution.

Exercise 5.12: Suppose X is a random variable taking values $1, 2$ and 3 and that $P(X = k) = p_k$, $k = 1, 2, 3$. Find the distribution $p = (p_1, p_2, p_3)$ having mean $\mu = 2.3$ for which the Euclidean norm $||p||_2$ is a minimum. Compare this with the maximum entropy distribution with mean $\mu = 2.3$. □

Exercise 5.13: Suppose X is a discrete random variable taking values x_i with probability p_i, $i = 1, ..., n$. Suppose further that for m given functions f_j the expected values

$$\mu_j = \sum_{i=1}^{n} p_i f_j(x_i)$$

are known. Show that the probability distribution with maximum entropy satisfying these conditions is given by

$$p_i = \exp(\sum_{j=1}^{m} \lambda_j f_j(x_i))/Z(\lambda)$$

where $Z(\lambda) = \sum_{i=1}^{n} \sum_{j=1}^{m} \lambda_j f_j(x_i)$ and the Lagrange multipliers are the solutions of the system

$$\frac{\partial}{\partial \lambda_j} \ln Z(\lambda) = \mu_j . \ \Box$$

In attempting to solve inverse problems we try to incorporate as much prior information as possible about the solution in the reconstruction process. The maximum entropy method as presented so far assumes as little as possible about the unknown distribution. However the entropy definition may be easily modified to include consideration of a "prior" distribution. The key to this modification is Shannon's inequality:

$$\sum_{i=1}^{n} p_i \ln \frac{p_i}{q_i} \geq 0 \tag{5.44}$$

where $(p_1, ..., p_n)$ and $(q_1, ..., q_n)$ are given probability distributions (we assume that $q_i > 0$). This result follows immediately from Jensen's inequality

$$E[\psi(X)] \geq \psi(E[X])$$

where E is the expectation operator and ψ is a convex function. In fact if we let X be the random variable that takes on the value p_i/q_i with probability q_i, then setting $\psi(x) = x \ln x$ we have

$$\sum_{i=1}^{n} q_i \frac{p_i}{q_i} \ln(\frac{p_i}{q_i}) \geq (\sum_{i=1}^{n} p_i) \ln(\sum_{i=1}^{n} p_i) = 0,$$

giving (5.44).

The modified entropy function, given a prior distribution $(q_1, ..., q_n)$ is

$$H(p_1, ..., p_n) = -\sum_{i=1}^{n} p_i \ln(\frac{p_i}{q_i}). \tag{5.45}$$

From (5.44) we see that this function is a maximum exactly when $p_i = q_i$ for each i. Therefore, without additional constraints, the maximum entropy distribution is the prior distribution. Also note that our earlier notion of entropy simply took the prior to be the uniform distribution. The more general form of the maximum entropy method then consists of finding a distribution $(p_1, ..., p_n)$ which maximizes (5.45) for a given prior and satisfies certain additional constraints dictated by measurements and observations.

The maximum entropy idea is by no means limited to discrete distributions. A simple case of the method applied to continuous distribution is given in the following exercise.

Exercise 5.14: Suppose that $p(t)$ represents a population a time $t \in [a, b]$ and for simplicity suppose that units are chosen so that

$$\int_a^b p(t)dt = 1.$$

Show that the population distribution $p(t)$ whose entropy

$$-\int_a^b p(t)\ln p(t)dt$$

is a maximum given by the common exponential growth model $p(t) = Ce^{kt}$, for some positive constants C and k. □

Recently, the maximum entropy idea has been used to regularize solutions of integral equations of the first kind. As in Tikhonov regularization, the idea is to seek a function which combines the features of a least squares solution with the regularity of an additional constraint by minimizing an augmented least squares functional. In the Tikhonov theory the regularizing term has the job of damping some norm of the solution, while in maximum entropy regularization the goal is to choose an approximate solution that has large entropy, or equivalently, small *negative entropy*

$$\nu(p) = \int_a^b p(t)\ln p(t)dt.$$

In attempting to approximate a nonnegative maximum entropy solution of

$$(Kx)(s) = \int_a^b k(s,t)x(t)dt = g(s)$$

one minimizes the functional

$$||Kx - g||^2 + \alpha\nu(x) \tag{5.46}$$

where

$$\nu(x) = \int_a^b x(t)\ln x(t)dt$$

and $\alpha > 0$ is a regularization parameter. This is a much costlier procedure than Tikhonov regularization because the minimization of (5.46) requires the solution of a *nonlinear* problem in contrast to the *linear* problem which must be solved to obtain the Tikhonov approximation.

5.6 The Backus-Gilbert Method

Is it possible to form a reasoned estimate of inaccessible values of a function from a few indirect measurements? This is of course at the heart of inverse theory and the method of Backus and Gilbert is designed to provide such estimates.

To illustrate the method, we consider a simple example. Imagine a solid impenetrable ball of radius 1 having variable density $\rho(r)$ depending only on the distance r from the center. Suppose the mass μ_1 of the ball can be measured. Then μ_1 is the value of a linear functional g_1 of ρ given by

$$\begin{aligned} \mu_1 &= \int_0^1 \int_0^{2\pi} \int_0^{\pi} \rho(r) r^2 \sin\phi \, d\phi d\theta dr \\ &= 4\pi \int_0^1 \rho(r) r^2 dr \\ &= <\rho, g_1>, \text{ where } g_1(r) = 4\pi r^2. \end{aligned}$$

Furthermore, suppose that the moment of inertia μ_2 of the ball about an axis through its center is known:

$$\begin{aligned} \mu_2 &= \int_0^1 \int_0^{2\pi} \int_0^{\pi} \rho(r) r^3 \sin^3\phi \, d\phi d\theta dr \\ &= \frac{8}{3}\pi \int_0^1 \rho(r) r^3 dr \\ &= <\rho, g_2>, \text{ where } g_2(r) = \frac{8}{3}\pi r^3. \end{aligned}$$

Given only the numbers μ_1 and μ_2, it is clearly impossible to determine the density $\rho(r)$ for $0 \le r \le 1$. But is it possible to make a reasonable estimate of the values of $\rho(r)$? To be definite, suppose we wish to estimate $\rho(.5)$, that is, the density half way between the center and the surface of the ball.

It would seem that the best we could hope for is to estimate some averaged value of ρ, say

$$\int_0^1 A(r)\rho(r)dr$$

where A is some averaging kernel satisfying

$$\int_0^1 A(r)dr = 1. \tag{5.47}$$

If we want to estimate $\rho(.5)$, then we would like

$$\rho(.5) \approx \int_0^1 A(r)\rho(r)dr \tag{5.48}$$

and hence we would like to "shape" the kernel $A(r)$ like the delta function $\delta(r-.5)$. In particular, we would like $A(r)$ to "peak" at $r = .5$. One way to arrange this peakedness is to force the quantity

$$\int_0^1 (A(r))^2 (r-.5)^2 dr \tag{5.49}$$

to be small. But how should $A(r)$ be formed? The only information we have is $\mu_1 =< \rho, g_1 >$ and $\mu_2 =< \rho, g_2 >$ and hence it is not unreasonable to take an estimate of the form

$$\begin{aligned} \rho(.5) &\approx a_1\mu_1 + a_2\mu_2 \\ &= < \rho, a_1 g_1 + a_2 g_2 > \\ &= \int_0^1 (a_1 g_1(r) + a_2 g_2(r))\rho(r)dr. \end{aligned} \tag{5.50}$$

In view of (5.48), we would then take

$$A(r) = a_1 g_1(r) + a_2 g_2(r)$$

where a_1 and a_2 are chosen so that, to satisfy (5.47),

$$a_1 \int_0^1 g_1(r)dr + a_2 \int_0^1 g_2(r)dr = 1. \tag{5.51}$$

To accomplish the peakedness criterion (5.49) we require that

$$\int_0^1 [a_1 g_1(r) + a_2 g_2(r)]^2 (r - .5)^2 dr = \min. \tag{5.52}$$

Therefore the unknown coefficients a_1 and a_2 are required to minimize the quadratic functional (5.52) while satisfying the linear constraint (5.51). This constrained optimization problem may be routinely handled by the Lagrange multiplier method yielding the coefficients a_1 and a_2. The Backus-Gilbert estimate of $\rho(.5)$ is then given by (5.51).

Exercise 5.15: Consider a circular disk of radius 1 and density distribution $\rho(r) = \frac{1}{\pi}(1 - r/2)$, $0 \leq r \leq 1$. The mass of the disk is then

$$\mu_1 = 2\pi \int_0^1 \rho(r) r dr = \frac{3}{2}$$

and the moment of inertia about a perpendicular axis through the center of the disk is

$$\mu_2 = 2\pi \int_0^1 \rho(r) r^3 dr = .3.$$

Use these values μ_1 and μ_2 to estimate $\rho(.5)$ by the Backus-Gilbert method and compare with the true value. Sketch the graph of the kernel $A(r)$. □

We now consider the Backus-Gilbert method in a bit more generality. Suppose n measurements, $\mu_1, \ldots, \mu_n$ are available which represent values of linearly independent functionals $g_1, \ldots, g_n$ on an unknown function ρ. The idea is to estimate $\rho(s)$ by a linear combination

$$a_1(s)\mu_1 + ... + a_n(s)\mu_n \tag{5.53}$$

by appropriately shaping the coefficients $a_1(s), ..., a_n(s)$. We then have

$$\begin{aligned} \rho(s) &\approx \sum_{j=1}^{n} a_j(s)\mu_j \\ &= \sum_{j=1}^{n} a_j(s) < \rho, g_j > \\ &= \int_0^1 A(s,r)\rho(r)dr \end{aligned} \tag{5.54}$$

where

$$A(s,r) = \sum_{j=1}^{n} a_j(s) g_j(r). \tag{5.55}$$

We require that

$$\int_0^1 A(s,r)dr = 1 \tag{5.56}$$

and that the peakedness condition

$$\int_0^1 A(s,r)^2(s-r)^2 dr = \min \tag{5.57}$$

is satisfied. Let us denote by $\rho_n(s)$ the estimate given by (5.55), that is,

$$\rho_n(s) = (a(s), \mu) \tag{5.58}$$

where $(\cdot,\cdot)$ is the euclidean inner product on $\mathbf{R}^n$, $\mu = [\mu_1, ..., \mu_n]^T$ and $a(s) = [a_1(s), ..., a_n(s)]^T$. Furthermore, if we define the inner product $< \cdot, \cdot >_s$ by

$$< f, h >_s = \int_0^1 f(r)h(r)(s-r)^2 dr \tag{5.59}$$

then we see from (5.57) that the peakedness condition is

$$\begin{aligned} \int_0^1 (A(s,r))^2(s-r)^2 dr &= < \sum_{i=1}^{n} a_i(s) g_i, \sum_{j=1}^{n} a_j(s) g_j >_s \\ &= (Ga(s), a(s)) = \min \end{aligned}$$

where $G = [< g_i, g_j >_s]$. The condition (5.56) is then

$$(a(s), \sigma) = 1$$

where $\sigma = [\sigma_1, ..., \sigma_n]^T$ with $\sigma_k =< 1, g_k >$.

Therefore, the coefficients $a(s) = [a_1(s), ..., a_n(s)]^T$ of the Backus-Gilbert method satisfy:

$$(Ga(s), a(s)) = \text{minimum}$$
$$\text{subject to: } (a(s), \sigma) = 1.$$

This is a standard quadratic minimization problem in $\mathbf{R}^n$ with linear constraints. The Lagrange multiplier method then provides a number λ such that

$$Ga(s) = \lambda\sigma \text{ and } (Ga(s), a(s)) = \lambda.$$

Therefore, we finally have

$$a(s) = G^{-1}\sigma/(G^{-1}\sigma, \sigma)$$

where $\sigma = [< 1, g_1 >, ..., < 1, g_n >]^T$ and $G = [< g_i, g_j >_s]$.

Exercise 5.16: For any given $s \in [0,1]$, the peakedness condition (5.57) is meant to guarantee that $\int_0^1 A(s,r)\rho(r)dr$ approximates $\rho(s)$ well. In this exercise we make this idea more precise. Consider the space H_s of all continuous functions on $[0,1]$ for which

$$||f||_s^2 = (f(s))^2 + \int_0^1 \left(\frac{f(s) - f(r)}{s - r}\right)^2 dr < \infty.$$

Show that $||\cdot||_s$ is a norm on H_s and that the linear functional that evaluates at s, $E_s\rho = \rho(s)$, is continuous on H_s. Define the linear functional A_s by

$$A_s(\rho) = \int_0^1 A(s,r)\rho(r)dr.$$

Show that the norm of the linear functional $A_s - E_s$ is

$$||A_s - E_s|| = \int_0^1 (A(s,r))^2(s-r)^2 dr. \; \square$$

5.7 ART

ART, or the algebraic reconstruction technique, is a simple iterative method for reconstructing pictures from projections. By a picture we mean a fixed two dimensional array of pixels with a nonnegative number (representing, for example, a density or weight) assigned to each pixel. In this context a projection simply consists of the sum of the values of selected pixels. We illustrate this with an example. Consider the arrangement in Figure 5.8. The pixels are ordered as indicated and the value of the pixels is specified by a vector

$$x = [x_1, x_2, ..., x_8]^T.$$

The six views $v^{(1)}, ..., v^{(6)}$ indicated give rise to the functionals

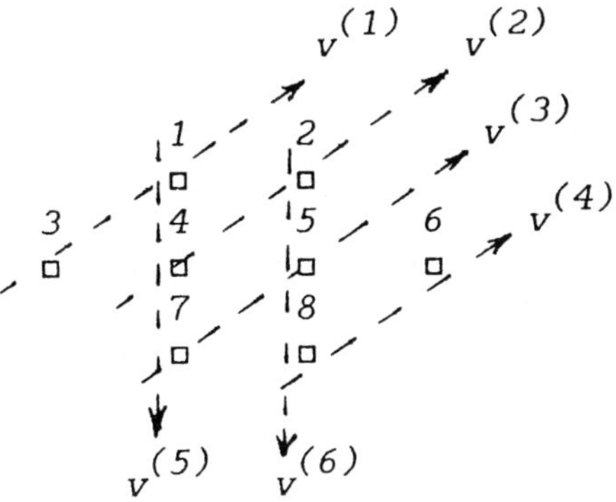

Figure 5.8

$$
\begin{aligned}
x_1 + x_3 &= (v^{(1)}, x) \\
x_2 + x_4 &= (v^{(2)}, x) \\
&\vdots \\
x_2 + x_5 + x_8 &= (v^{(6)}, x)
\end{aligned}
$$

where $(\cdot, \cdot)$ is the euclidean inner product and

$$
\begin{aligned}
v^{(1)} &= [1,0,1,0,0,0,0,0]^T \\
v^{(2)} &= [0,1,0,1,0,0,0,0]^T \\
&\vdots \\
v^{(8)} &= [0,1,0,0,1,0,0,1]^T
\end{aligned}
$$

For example, the weights $x = [1, 1, 1.5, 2, 2.5, 1, 1, 2]^T$ would give the results

$$
\begin{aligned}
(v^{(1)}, x) &= 2.5 \\
(v^{(2)}, x) &= 3 \\
&\vdots \\
(v^{(6)}, x) &= 5.5.
\end{aligned}
$$

The reconstruction problem consists of reconstructing the weights x given the views $v^{(j)}$ and the projections $(v^{(j)}, x)$. Depending on the number of pixels and the number of views, this problem could be underdetermined, as in the example above, or overdetermined (although this is unlikely in a practical situation).

Before presenting the basic ART algorithm, we establish some basic facts about projections onto hyperplanes. Consider the hyperplane

$$H = \{x :< v, x >= \mu\}$$

in an inner product space, where v is a given nonzero vector and μ is a given scalar. The vector of smallest norm H is $\mu v/||v||^2$ and it is easy to see that

$$H = \mu v/||v||^2 + v^{\perp} \qquad (5.60)$$

where $v^{\perp} = \{z :< v, z >= 0\}$. Since H is closed and convex, for each x there is a unique vector $Px \in H$ with

$$||x - Px|| = \min\{||x - y|| : y \in H\}.$$

It follows that the function

$$g(t) = ||x - (Px + tw)||^2$$

has a minimum at $t = 0$ for each w with $Px + tw \in H$. In view of (5.60), the minimum of $g(t)$ is achieved at $t = 0$ for each $w \in v^{\perp}$. Setting $g'(0) = 0$, we find that

$$< x - Px, w >= 0 \text{ for all } w \in v^{\perp}$$

and hence

$$x - Px \in v^{\perp\perp} = \{\alpha v : \alpha \in \mathbf{R}\}.$$

Therefore $Px = x - \alpha v$, for some scalar α. Since $Px \in H$, we find that

$$\mu =< v, Px >=< v, x > -\alpha||v||^2$$

that is,

$$Px = x + \frac{\mu - < v, x >}{||v||^2} v \qquad (5.61)$$

Exercise 5.17: Show that if P is the projection operator defined by (5.61), then for any $z \in H$,

$$||Py - z||^2 \leq ||y - z||^2 - (\mu - < v, y >)^2/||v||^2.$$

From this it follows immediately that $||Py - Px|| \leq ||y - x||$ for any x and y. We then say that the operator P is *nonexpansive*. Show, more generally, that if P is the projection operator onto a closed convex set in Hilbert space, then P is nonexpansive. □

A more general reconstruction problem may be described as follows. Given certain "view" vectors $v^{(1)}, v^{(2)}, ..., v^{(m)}$ and some scalars $\mu_1, \mu_2, ..., \mu_m$, find a vector x satisfying

$$< v^{(j)}, x >= \mu_j, \quad j = 1, ..., m. \tag{5.62}$$

If we denote by H_j the hyperplane determined by $v^{(j)}$ and μ_j, that is,

$$H_j = \{x :< v^{(j)}, x >= \mu_j\},$$

then we seek a vector $x \in C = \bigcap_{j=1}^m H_j$.

In its most primitive form, the algebraic reconstruction technique consists of successively projecting a given vector $x^{(0)}$ (we will take $x^{(0)} = 0$) on the hyperplanes H_j in a cyclic fashion. That is,

$$\begin{aligned} x^{(0)} &= 0 \\ x^{(1)} &= P_1 x^{(0)}, x^{(2)} = P_2 x^{(1)}, ..., x^{(m)} = P_m x^{(m-1)} \\ x^{(m+1)} &= P_1 x^{(m)}, \text{ etc.} \end{aligned}$$

where P_j is the projector onto the hyperplane H_j. To put it another way,

$$\begin{aligned} x^{(k+1)} &= P_j x^{(k)} \\ &= x^{(k)} - \frac{\mu_j - < v^{(j)}, x^{(k)} >}{\|v^{(j)}\|^2} v^{(j)}, \quad k = 0, 1, 2, ... \end{aligned} \tag{5.63}$$

where $j = k(\text{mod}m) + 1$.

A fairly straightforward argument, based on Exercise 5.17, establishes the convergence of the ART method. Note that if

$$\bar{x} \in C = \bigcap_{j=1}^{m} H_j,$$

then by (5.64) and the exercise, we have

$$0 \leq \|x^{(k+1)} - \bar{x}\| \leq \|x^{(k)} - \bar{x}\|$$

and hence $\|x^{(k)} - \bar{x}\|$ converges as $k \to \infty$. By Exercise 5.17, we then have

$$\lim_{k \to \infty} (\mu_j - < v^{(j)}, x^{(k)} >) = 0 \tag{5.64}$$

where $j = k(\text{mod}m) + 1$. Since $\{x^{(k)}\}$ is a bounded sequence in a finite dimensional space, it has a cluster point x. We will show that $x \in C$ and $x^{(k)} \to x$ as $k \to \infty$. From (5.64) and (5.64) we find that

$$\|x^{(k+1)} - x^{(k)}\| \to 0 \tag{5.65}$$

and hence

$$\|x^{(k+j)} - x^{(k)}\| \to 0 \text{ as } k \to \infty \tag{5.66}$$

for $j = 1, 2, ..., m$. Now, if $\{x^{(k_n)}\}$ is a subsequence with

$$x^{(k_n)} \to x \text{ as } n \to \infty,$$

then, from (5.66),

$$x^{(k_n+j)} \to x \text{ as } n \to \infty \text{ for } j = 1, 2, ..., m.$$

From this it follows that

$$x \in C = \bigcap_{j=1}^{m} H_j.$$

Finally, from (5.65) we obtain

$$x^{(k)} \to x \text{ as } k \to \infty.$$

We can say a bit more about the particular vector to which the ART method converges. Note that a solution x of (5.62) is just a solution of the $m \times n$ linear system

$$Vx = b$$

where V is the matrix whose rows are the vectors $v^{(j)T}$ and b is the m-vector $[\mu_1, ..., \mu_m]^T$. Now, it is easy to see that since $x^{(0)} = 0$, $x^{(k)}$ is in the space $R(V^T) = N(V)^\perp$, for each k and hence $x \in N(V)^\perp$, that is, assuming that $c \neq \emptyset$, $x = V^\dagger b$, the minimum norm solution of (5.62). More generally, for any $x^{(0)}$, the ART method will converge to the solution of (5.62) which is nearest to $x^{(0)}$. In this way, a priori information, in the form of $x^{(0)}$, can be introduced into the reconstruction algorithm.

Exercise 5.18: In picture reconstruction problems we would presumably seek a solution x which has only nonnegative components. Show that if we set

$$x_+^{(k)} = \max\{0, x^{(k)}\}$$

for each k, then this amounts, at each iteration, to projecting onto an additional convex set (the nonnegative vectors). Show that if $x_+^{(0)} = 0$, then $x_+^{(k)}$ converges to the nonnegative solution (assuming such exists) of (5.62) with minimum norm. □

Exercise 5.19: Write a computer program implementing the ART method and test it on the example that was introduced at the beginning of this section. □

5.8 Ouput Least Squares

Most of the inverse problems treated in Chapter 2 are linear problems. However, problems involving identification of coefficients in differential equations, even *linear* differential equations, lead to nonlinear inverse problems. To illustrate this, consider an extremely simple coefficient determination problem, namely the problem of determining the *constant* coefficient a in the initial value problem

$$y' - ay = f, \;\; y(0) = 1, \tag{5.67}$$

from knowledge of the solution y. The nonlinear dependence of a on y is apparent when (5.67) is solved for a, but we would like to formulate the inverse problem in an implicit way. Given the initial condition, the forcing function f, and the coefficient a, the direct problem of finding y is a standard elementary exercise yielding:

$$y(t) = e^{at} \int_0^t e^{-as} f(s) ds + 1.$$

Operationally, we would like to interpret this in terms of a coefficient-to-solution operator

$$F(a) = y \tag{5.68}$$

where the operator F is defined by

$$(F(a))(t) = e^{at} \int_0^t e^{-as} f(s) ds + 1. \tag{5.69}$$

The inverse problem of determining a from y in the linear problem (5.67) is now a nonlinear problem in a as reflected in (5.69). It is also clear that if a is a constant, then the inverse problem (5.68) is severely overdetermined in that (5.68) specifies a condition this single constant must satisfy for every t.

Exercise 5.20: Suppose that f is continuous and $f(s) \geq c > 0$ and let $d = \int_0^1 f(s)ds$. Consider the inverse problem of determining the constant $a > 0$ from a single measurement, $y(1)$, of the solution. Suppose μ is a measured value of $y(1)$. Show that if $\mu < d$, then the problem $F(a) = \mu$ has no solution, while if $d \leq \mu$, then $F(a) = \mu$ has a unique solution. □

Exercise 5.21: Consider the initial value problem $y' - ay = 1$, $y(0) = 2$, where a is a variable coefficient. Let $y_n(t) = \frac{1}{n} \sin nt + 2$, $n = 1, 2, \ldots$. Show that there is a unique continuous coefficient $a_n(t)$ for which $y_n(t)$ is the solution of the initial value problem. Also show that $y_n \to 2$ uniformly in t, but $\{a_n(t)\}$ converges only for $t = 0$. □

We close with a few general remarks on methods for the nonlinear inverse problem

$$F(a) = y. \tag{5.70}$$

An often used method for solving such problems is called *output least squares*. The goal of this method is to find a least squares solution a^* of (5.70). That is, if F is defined on some suitable class $\mathcal{D}(F)$ of parameters, one seeks a $a^* \in \mathcal{D}(F)$ such that

$$\|F(a^*) - y\| = \inf\{\|F(a) - y\| : \; a \in \mathcal{D}(F)\}$$

assuming that such a function a^* exists. As always, the function y is known only to within a certain tolerance δ, that is, an approximation y^δ satisfying

$$\|y - y^\delta\| \leq \delta$$

is known and one therefore seeks an a^* minimizing

$$\|F(a^*) - y^\delta\|.$$

The idea of the method is to start with an admissible $a_0 \in \mathcal{D}(F)$, solve the *forward* problem

$$F(a_0) = y_0$$

and then update a_0 depending on how well y_0 matches y^δ (in the petroleum industry, the idea is called *history matching* because y^δ typically represents production history at a given set of wells). The updating procedure varies, but one standard technique is to assume that F is Fréchet differentiable and employ the linearization

$$F(a_0 + h) = F(a_0) + F'(a_0)h + r(a_0; h) \tag{5.71}$$

where the linear operator $F'(a_0)$ is the Fréchet derivative of F at a_0, that is, the remainder $r(a_0; h)$ satisfies

$$\|r(a_0; h)\| = o(\|h\|).$$

Since $F(a_0) = y_0$ and we wish to choose an update h to a_0 so that $F(a_0 + h) = y^\delta$, the remainder term in (5.71) is dropped and the update h is taken as the generalized solution of the linear operator equation

$$F'(a_0)h = y^\delta - y_0.$$

Of course, this equation is generally ill-posed and hence Tikhonov regularization can be employed and h is taken as the solution of

$$(F'(a_0)^* F'(a_0) + \alpha I)h = F'(a_0)^*(y^\delta - y_0)$$

(in this context, the use of Tikhonov regularization is called the Levenberg-Marquardt method in the optimization literature). Once the correction h is determined, the estimate a_0 of the coefficient is updated to $a_0 + h$ and the process is repeated until the output $F(a_0)$ is a sufficiently close match to the measured data y^δ. The output least squares method then consists of cycles of the following steps: forward-solve, linearize, regularize and update. The success of the method clearly depends on the availability of accurate direct problem solvers for the forward-solve step and good linearizations.

Exercise 5.22: Let F be the coefficient-to-solution operator, $F(a) = y$, for the initial value problem $y' - ay = 1$, $y(0) = 2$. Show that $F'(a)$ is the linear integral operator

$$(F'(a)h)(t) = \int_0^t k(t,\tau)h(\tau)d\tau$$

where $k(t,\tau) = e^{\mu(t)}[\int_0^\tau e^{-\mu(s)}ds + 2]$ and $\mu(t) = \int_0^t a(x)dx$. □

In the output least squares method the problem is linearized and then regularized. An alternative is to regularize first. This approach is called ***penalized least squares*** and can be regarded as Tikhonov regularization of a nonlinear problem. In penalized least squares one seeks a minimum in $\mathcal{D}(F)$ of the functional

$$\|F(a) - y^\delta\|^2 + \alpha\|a\|^2 \tag{5.72}$$

where $\alpha > 0$ is a regularization parameter. The minimization of the functional (5.72) requires a discretization scheme and the use of appropriate optimization software. However, the existence of minimizers of (5.72) is assured under mild assumptions on F.

Exercise 5.23: Suppose that $F : \mathcal{D}(F) \subseteq H_1 \to H_2$ is weakly closed, that is, the graph $\mathcal{G}(F)$ of F is weakly closed in $H_1 \times H_2$. Show that (5.72) has a (not necessarily unique) minimum in $\mathcal{D}(F)$. □

One can also show under relatively mild assumptions that, if (5.70) has a least squares solution, then for any sequences $\delta_n \to 0$, $\alpha_n \to 0$ with δ_n/α_n bounded, then any sequence of minimizers of

$$\|F(a) - y^{\delta_n}\|^2 + \alpha_n\|a\|^2,$$

where $\|y - y^{\delta_n}\| \le \delta_n$, has a strongly convergent subsequence and the limit of every such strongly convergent subsequence is a least squares solution of (5.70) (see [BE]). It is in this sense that the convergence theory of Tikhonov regularization carries over to the nonlinear problem (5.70).

5.9 Bibliographic Notes

Tikhonov's paper [Ti] is usually taken as the origin of the theory of regularization, however, it was preceded by the paper of Phillips [Ph] which contains the essential idea of regularization as well as an early version of the discrepancy principle ([Ck] is another early paper that is not often recognized). More recent monographs in English on the subject include [TA], [G], [Gro2], [H], [LRS] and [Bau] (see also the German monograph [Lo]).

The Landweber iteration method is developed in [La] (see [AMD] for applications and [Hnk] for some modern developments). The major drawback of the method is its slow convergence. Multigrid ideas are applied to the method to speed its convergence in [ZR] and [Kg1]. For information on the conjugate gradient method and variants, see [KN], [Brk], [Lo1] and [Kg]. [BFMW] and [Ha1] are early efforts at applying the TSVD method to solve Fredholm integral equations of the first kind numerically. For details on more recent work see [V1], [Vo] and [Hns]. The plots in this chapter were produced with the help of [Hns1].

There is a huge literature on the maximum entropy method. [Ri] is a good introduction. For the application of the maximum entropy idea to Fredholm integral equations of the first kind, see [KS], [AH], [E], and [EL]. The Backus-Gilbert method [BG] is a standard tool of the geophysics community. For more information along the lines of our treatment, see [SB], [R] and [Maa].

[CH] (see also [Her]) is an excellent survey of iterative reconstruction methods, including ART. The proof of the convergence of the ART algorithm given here follows [Mar 3] (see also [Tru]). For extensions to successive projections on closed convex sets, see [Yo]. [BK], [Vo1] and [CER] contain good surveys of methods for nonlinear ill-posed problems. [Zhu] is an interesting news article on history matching in the petroleum industry. A theoretical analysis of the penalized least squares method can be found in [BE].

6 An Annotated Bibliography on Inverse Problems

Search the scriptures
John v : 39

[**AMD**] J.B. Abbiss, C. deMol and H. Dhadwal, Regularized iterative and noniterative procedures for object restoration from experimental data, *Optica Acta* 30(1983), 107-124.

An integral equation of the first kind arising in optics is solved by Tikhonov regularization and by an iterative method. Numerical illustrations of the procedures applied to real data are also provided.

[**AS**] A. Abramowitz and I. Stegun (Eds.), *Handbook of Mathematical Functions*, U.S. Department of Commerce, Washington, D.C., 1964.

A valuable reference containing a wealth of information on special functions and much more.

[**Al**] H. Allison, Inverse unstable problems and some of their applications, *The Mathematical Scientist* 4(1979), 9-30.

A first rate expository article. The author shows that inverse problems are widespread and physically meaningful and that classical notions of solution and conventional numerical methods are not generally applicable to such problems.

[**AH**] U. Amato and W. Hughes, Maximum entropy regularization of Fredholm integral equations of the first kind, *Inverse Problems* 7(1991), 793-808.

A well-posed problem, the solution of which approximates a nonnegative solution of a linear Fredholm integral equation of the first kind, can be formulated by minimizing a Tikhonov-like functional in which the usual quadratic regularization term is replaced by a negative "entropy" term. It is shown that under appropriate circumstances this procedure gives an approximation method that is regular in the sense of Tikhonov.

[**A**] D.H. Anderson, *Compartmental Modeling and Tracer Kinetics*, LNB 50, Springer-Verlag, New York, 1983.

A well written treatment, with lots of examples, of the tracer kinetics problem and related issues.

[**AHL**] R.S. Anderssen, F.R. deHoog and M.A. Lukas (Eds.), *The Application and Numerical Solution of Integral Equations*, Sijthoff and Noordhoff, Alphen an den Rijn, 1980.

Proceedings of a seminar at Australian National University. The papers by Anderssen (Abel equations), deHoog (Fredholm equations of the first kind) and Lukas (regularization) are particularly recommended.

[AD] R.S. Anderssen and C.R. Dietrich, The inverse problem of aquifer transmissivity identification, in [EG], pp. 19-28.

This paper surveys some of the methods proposed for identifying the transmissivity coefficient in an aquifer and examines the use of the linear functional strategy for estimating a piecewise constant approximation to the transmissivity coefficient.

[AN] R.S. Anderssen and G.N. Newsam (Eds.), *Special Program on Inverse Problems*, Proceedings of the Centre for Mathematical Analysis, vol 17, Canberra, Australia, 1988.

A collection of papers from a special year on inverse problems. Topics include numerical differentiation, regularization, algebraic inverse problems, inverse vibration problems, and identification problems for ordinary differential equations.

[Ang] G. Anger, *Inverse Problems in Differential Equations*, Plenum Press, New York, 1990.

A monograph on inverse source and coefficient problems for differential equations stressing inverse problems in potential theory. The book contains an interesting historical postscript and an extensive bibliography, particularly on eastern European work.

[An] Anonymous (Ed), *Computed Tomography*, Proceedings of Symposia in Applied Mathematics, vol. 27, American Mathematical Society, Providence, 1983.

Notes from an AMS short course on tomography. The historical paper by Cormack is particularly interesting. The notes also contain a reprint of Radon's 1917 paper.

[BG] G.E. Backus and J.F. Gilbert, The resolving power of gross earth data, *Geophysical Journal of the Royal Astronomical Society* 16(1968), 169-205.

The original paper in which the famous Backus-Gilbert method for estimating values of an unknown function by using a limited number of measured functional values is investigated.

[B] C.T.H. Baker, *The Numerical Treatment of Integral Equations*, Clarendon Press, Oxford, 1977.

An encyclopedic source of information on numerical methods for integral equations, including methods for integral equations of the first kind.

[BFMW] C.T.H. Baker, L. Fox, D. Mayers and K. Wright, Numerical solution of Fredholm integral equations of the first kind, *The Computer Journal* 7(1964), 141-147.

An early paper investigating the use of the truncated singular value decomposition for numerical solution of integral equations of the first kind.

[Bak] A.B. Bakushinskii, Iterative method for the solution of nonlinear ill-posed problems and their applications, in *Ill-posed Problems in Natural Sciences* (A.S. Leonov, et al., Eds.), VSP Science Publishers, Utrecht, 1992, pp. 13-17.

A very succinct but clearly written account of regularized iterative methods for ill-posed nonlinear operator equations. The case of monotone operators and regularized Newton-like methods for operators satisfying certain differentiability conditions are considered.

[BK] H.T. Banks and K. Kunisch, *Estimation Techniques for Distributed Parameter Systems*, Birkhäuser, Boston, 1989

A research monograph on approximational and computational aspects of inverse problems for infinite dimensional systems, specifically coefficient identification problems in partial differential equations. The mathematical concepts and techniques are introduced and motivated by models in biology and mechanics.

[Bau] J. Baumeister, *Stable Solution of Inverse Problems*, Vieweg, Braunschweig, 1987.

Lecture notes for an advanced course on inverse and ill-posed problems. Chapter topics: ill-posed problems, regularization, SVD, Tikhonov methods, discretization, least squares problems, convolution, the final value problem, and parameter identification.

[BBC] J.V. Beck, B. Blackwell, and C. St. Clair, Jr., *Inverse Heat Conduction: Ill-posed Problems*, Wiley, New York, 1985.

The topic of this book is the estimation of the surface heat flux history of a body from measurements taken in the interior of the body. Analytical and numerical methods are discussed and the ill-posed nature of the problem is emphasized. The book is intended as a text and contains many exercises, examples and references.

[BRRW] J. Bednar, R. Redner, E. Robinson, and A. Weglein (Eds.), *Conference on Inverse Scattering: Theory and Applications*, SIAM, Philadelphia, 1983.

This volume contains a long expository article on inverse scattering by R. Newton, followed by a number of shorter research papers on inverse scattering and related theoretical issues.

[Be] R.J. Bell, *Introduction to Fourier Transform Spectroscopy*, Academic Press, New York, 1972.

An excellent source for the basics of the Fourier transform and its applications in spectroscopy.

[Bel] R. Bellman, *Mathematical Methods in Medicine*, World Scientific, Singapore, 1983.

The emphasis of this book is compartmental analysis and the motivating applications in pharmacokinetics. Model building, numerical methods, optimal dosages, tumor detection and radiotherapy are discussed.

[BIG] A. Ben-Israel and T. Greville, *Generalized Inverses: Theory and Applications*, Wiley, New York, 1974.

A standard text on generalized inverses of matrices which also includes a chapter on generalized inverses of operators in Hilbert space. The book contains a wealth of exercises and examples.

[BA] B. Berkstresser, S. El-Asfouri, J. McConnell, and B. McInnis, Identification techniques for the renal function, *Mathematical Biosciences* 44(1979), 157-165.

A study of the use of compartmental analysis to identify parameters related to renal blood flow.

[Ber] M. Bertero, Linear inverse and ill-posed problems, *Advances in Electronics and Electron Physics* 75(1989), 2-120.

A detailed and well-written survey, from a physicist's perspective, of the theory of linear ill-posed operator equations. The monograph has chapters on linear inverse problems, linear inverse problems with discrete data, generalized solutions, regularization theory for ill-posed problems, inverse problems and information theory, and an extensive bibliography.

[BDV] M. Bertero, C. DeMol and G. Viano, The stability of inverse problems, in *Inverse Scattering Problems in Optics* (H. Baltes, Ed.), Topics in Current Physics, vol.20, Springer-Verlag, New York, 1980, pp. 161-214.

A survey of general mathematical techniques for ill-posed problems in physics. The fact that development of adequate stable computational methods requires certain prior knowledge of solutions (global bounds, smoothness, convexity, statistical properties, etc.) is stressed.

[BE] A. Binder, H.W. Engl, et al., Weakly closed nonlinear operators and parameter identification in parabolic equations by Tikhonov regularization, *Institutsbericht No. 444*, University of Linz, 1991.

A convergence analysis of the penalized least squares method for nonlinear ill-posed operator equations in Hilbert space with applications to coefficient identification in parabolic partial differential equations.

[Bl] F. Bloom, *Ill-posed Problems for Integrodifferential Equations in Mechanics and Electromagnetic Theory*, SIAM, Philadelphia, 1981.

The emphasis of this monograph is stability and other qualitative properties of solutions of some ill-posed nonlinear evolution problems. The primary techniques are logarithmic convexity and concavity arguments.

[**Bo**] N.N. Bojarski, Inverse black body radiation, *IEEE Transactions on Antennas and Propagation*, 30(1982), 778-780.

The inverse black body radiation problem is introduced in this note. An iterative solution method involving the computation of an inverse Laplace transform is suggested.

[**Bol**] B.A. Bolt, What can inverse theory do for applied mathematics and the sciences?, *Australian Mathematical Society Gazette* 7(1980), 69-78.

The text of a K.E. Bullen Lecture delivered at the University of Sydney. A plea is made for the teaching of inverse problems in the undergraduate curriculum. Inverse problems relating to the earth's density, a particular interest of K.E. Bullen, are used as vehicles to introduce inverse theory.

[**Brw**] R.N. Bracewell, Strip integration in radioastronomy, *Australian Journal of Physics* 9(1956), 198-217.

An early paper on tomography containing illustrations of Abel integral equations in inverse problems in astronomy.

[**Brk**] H. Brakhage, On ill-posed problems and the method of conjugate gradients, in [EG], pp.165-176

Aspects of the theory of orthogonal polynomials are used to establish rates of convergence of the conjugate gradient method for bounded linear operator equations. The case of erroneous data is also considered.

[**Br**] W.C. Brenke, An application of Abel's integral equation, *American Mathematical Monthly* 29(1922), 58-60.

A treatment of the weir notch problem in irrigation.

[**BB**] K.E. Bullen and B. Bolt, *An Introduction to the Theory of Seismology*, 4th Ed., Cambridge University Press, Cambridge, 1985.

A posthumous revision of Bullen's 1947 classic. Contains all necessary background material on seismology and a discussion of the inverse problem of seismic travel times and other geophysical inverse problems.

[**Bu**] R. Burridge, *Some Mathematical Questions in Seismology*, New York University, 1975.

Lecture notes from a course at the Courant Institute of Mathematical Sciences in 1974-75. The ninth chapter, on the Backus-Gilbert theory of the geophysical inverse problem, is particularly recommended.

[CB] H. Cabayan and G. Belford, On computing a stable least squares solution to the inverse problem for a planar Newtonian potential, *SIAM Journal of Applied Mathematics* 20(1971), 57-61.

The problem of determining the shape of an unknown planar gravitating body from external potential measurements is considered. The nonlinear Fredholm integral equation of the first kind is solved by Tikhonov regularization.

[Cp] S. Campi, An inverse problem related to the travel time of seismic waves, *Bolletino della Unione Matematica Italliana* 17(Ser.B)(1980), 661-674.

The problem is to determine the refractive index of the earth from measurements of the travel time of seismic waves from one point of the surface to another. The author casts the problem as a nonlinear integral equation and obtains existence and stability results.

[Ca] J. Cannon, *The One-dimensional Heat Equation*, Addison-Wesley, Menlo Park, 1984.

A comprehensive account of the classical theory of the one dimensional heat equation. The book is a particularly valuable reference for heat equations with source terms. It includes a hugh bibliography that is interestingly divided into time periods in the development of the subject.

[CaH] J. Cannon and U. Hornung (Eds.), *Inverse Problems*, Birkhäuser, Basel, 1986.

Proceedings of a conference held at the Mathematics Research Institute, Oberwolfach, Germany in 1986. Topics include inverse source problems, the linear functional strategy for inverse problems, inverse scattering theory, stereology, inverse potential theory, inverse problems for parabolic equations and approximate methods for abstract ill-posed problems.

[CH] Y. Censor and G.T. Herman, On some optimization techniques in image reconstruction from projections, *Applied Numerical Mathematics* 3(1987), 365-391.

A very useful survey, with an extensive bibliography, of iterative optimization methods for approximating solutions of discretized versions of the problem of determining a function from certain of its line integrals.

[Ch] M.T. Chahine, Determination of the temperature profile in an atmosphere from its outgoing radiance, *Journal of the Optical Society of America* 58(No.12) (1968), 1634-1637.

An iterative method is developed for the radiative transfer equation of atmospheric temperature profiles.

[CMMA] S. Chakravarty, M. Menguc, D. Mackovski, and R. Altenkirch, Application of two inversion schemes to determine the absoption coefficient distribution in flames, *Proceedings of the ASME National Heat Transfer Conference,* Houston, 1988.

The basic Abel equation model of simplified tomography is applied to the problem of determining the structure of a flame.

[CL] N.-X. Chen and G.-Y. Li, Theoretical investigation of the inverse black body radiation problem, *IEEE Tansactions on Antennas and Propagation*, 38(No.8)(1990), 1287-1290.

Various methods for solving the inverse black body radiation problem are reviewed and a new method is suggested.

[Co] R. Coleman, Inverse problems, *Journal of Microscopy* 153(1989), 233-248.

A survey of linear inverse problems expressed as integral equations of the first kind and methods of regularization.

[Col] L. Colin (Ed.), *Mathematics of Profile Inversion*, NASA Technical Memorandum TM X-62-150, 1972.

This relatively hard-to-find source is a collection of papers, some quite readable, from an early workshop on atmospheric remote sensing held at NASA's Ames Research Center in Moffet Field, California.

[CER] D. Colton, R. Ewing and W. Rundell (Eds.), *Inverse Problems in Partial Differential Equations*, SIAM, Philadelphia, 1990.

Proceedings of an AMS-SIAM Summer Research Conference. Topics include inverse scattering, impedance imaging, complexity of ill-posed problems, approximation techniques, and the Dirichlet to Neumann map.

[CK] D. Colton and R. Kress, *Integral Equation Methods in Scattering Theory*, Wiley, New York, 1983.

A nice self-contained monograph on scattering theory for the Maxwell and Helmholtz equations. The necessary background material on integral equations is covered in the text and there is a chapter on improperly posed problems.

[CK2] D. Colton and R. Kress, *Inverse Acoustic and Electromagnetic Scattering Theory*, Springer-Verlag, New York, 1992.

This substantial extension and update of [CK] emphasizes the nonlinear and ill-posed nature of the inverse scattering problem. Topics include direct scattering for Helmholtz and Maxwell equations, ill-posed problems, inverse acoustic and elctromagnetic obstacle scattering, acoustic and electromagnetic waves in an inhomogeneous medium, and the inverse medium problem.

[Ck] B. Cook, Least structure solution of photonuclear yield functions, *Nuclear Instruments and Methods* 24(1963), 256-268.

An early, and often overlooked, paper applying (Tikhonov-) Phillips regularization to a Fredholm integral equation of the first kind.

[CKNS] J. Corones, G. Kristensson, P. Nelson and D. Seth (Eds.), *Invariant Imbedding and Inverse Problems,* SIAM, Philadelphia, 1992.

A collection of research papers dedicated to the memory of Robert Krueger. A central theme in transport theory is the determination of propagation characteristics from permitivity and conductivity profiles of an inhomogeneous medium. These papers treat the determination of such profiles from electromagnetic scattering data and related problems.

[CrB] I. Craig and J.Brown, *Inverse Problems in Astronomy,* Adam Hilger, Bristol, 1986.

Treats Fredholm and Volterra integral equations of the first kind occurring in astronomy. The physics is rough going, but the mathematics is comprehensible.

[De] A. Deepak (Ed.), *Inversion Methods in Atmospheric Remote Sounding,* Academic Press, New York, 1977.

A collection of twenty one papers on inverse problems in atmospheric sounding. Topics include radiative transfer, mathematical theory of inversion methods, and inversion methods in thermal, gaseous and aerosol atmospheres.

[DW] L.M. Delves and J. Walsh (Eds.), *Numerical Solution of Integral Equations,* Clarendon Press, Oxford, 1974.

A collection of expository articles on numerical methods for integral equations. Of particular interest are the chapter by Baker on Volterra equations of the first kind and the chapter by Miller on Fredholm equations of the first kind.

[DS] J.E. Dennis, Jr. and R.B. Schnabel, *Numerical Methods for Unconstrained Optimization and Nonlinear Equations,* Prentice-Hall, Englewood Cliffs, N.J., 1983.

An excellent source of theory and algorithms for systems of nonlinear equations.

[DH] P. Deuflhard and E. Hairer (Eds.), *Numerical Treatment of Inverse Problems in Differential and Integral Equations,* Birkhäuser, Basel, 1983.

A collection of research papers covering inverse initial value problems, boundary value problems and eigenvalue problems in ordinary differential equations, inverse problems in partial differential equations and Fredholm integral equations of the first kind.

[Di] P. Dive, Sur l'identité de deux corps possédent le même potentiel newtonien dans une région intérieure commune, *Comptes Rendus des Séances de* l' *Académie de Sciences (Paris)* 195(1932), 597-599.

The uniqueness of the interior inverse problem of gravitational potential is established for convex bodies in which the density is an analytic function of position.

[D] L.M. Dorman, The gravitational edge effect, *Journal of Geophysical Research,* 80(No.20)(1975), 2949-2950.

The problem of determining the density difference between two adjacent laterally uniform subterranean structures (such as at a continental margin) by measurements of the surface gravity anomoly is cast as a Fredholm integral equation of the first kind.

[Dr] K. Dressler, Inverse problems in linear transport theory, *European Journal of Mechanics,* B/*Fluids* 8(1989), 351-372.

Some inverse problems in neutron transport theory, which in some cases give rise to an ill-posed Fredholm equation of the first kind, are investigated.

[E] P.P.B. Eggermont, Maximum entropy regularization for Fredholm integral equations of the first kind , *preprint*, University of Delaware, 1991.

A fairly complete Tikhonov-like convergence program is carried out for entropy-regularized approximate solutions of Fredholm integral equations of the first kind.

[El] L. Eldèn, Modified equations for approximating the solution of a Cauchy problem for the heat equation, in [EG], pp. 345-350.

The idea is to regularize the inverse heat conduction problem by modifying the governing partial differential equation so as to change its type (from parabolic to hyperbolic).

[EG] H.W. Engl and C.W. Groetsch (Eds.), *Inverse and Ill-posed Problems*, Academic Press, Orlando, 1987.

Papers presented at the Alpine-U.S. Seminar on Inverse and Ill-posed Problems held in St. Wolfgang, Austria in June, 1986. Topics include the theory of ill-posed problems, regularization methods, tomography, inverse scattering, source detection, inverse heat problems, Cauchy problems, and parameter estimation.

[EKN] H.W. Engl, K. Kunisch and A. Neubauer, Convergence rates for Tikhonov regularization of nonlinear ill-posed problems, *Inverse Problems* 5(1989), 523-540.

Tikhonov regularization for nonlinear ill-posed problems is shown to be stable with respect to data errors and a convergence rate with respect to a bound for the data error is provided. The paper also contains an interesting discussion of the connection between the ill-posedness of a nonlinear problem and its linearization.

[EL] H.W. Engl and G. Landl, Convergence rates for maximum entropy regularization, *Report No. 445,* Institute for Mathematics, University of Linz, Austria, 1991.

Convergence rates for entropy-regularized approximations to the maximum entropy solution of a Fredholm integral equation of the first kind are proved by converting the entropy augmented least squares functional for the linear problem into a Tikhonov functional for an equivalent *nonlinear* problem.

[FW] V. Faber and G.M. Wing, The Abel integral equation, Report LA-11016-MS, Los Alamos National Laboratory, Los Alamos, New Mexico, 1987.

An example of a tomographic experiment leading to an Abel integral equation is provided and a numerical method based on the singular value decomposition is discussed. The results of several numerical experiments are reported.

[FWZ] V. Faber, G.M. Wing and J. Zahrt, Inverse problems, SVD, and Pseudo SVD, in *Invariant Imbedding and Inverse Problems* (J. Corones, et al., Eds.) SIAM, Philadelphia, 1992, pp. 230-240.

The singular value decomposition is discussed as a tool for optimizing the condition number of discrete approximations to an unknown kernel appearing in a Fredholm integral equation of the first kind.

[F] R.S. Falk, Approximation of inverse problems, in [CER], pp.5-14.

The author derives error estimates for a finite element method for approximating a distributed coefficient in an elliptic partial differential equation.

[FZ] A.L. Fymat and V. Zuev,(Eds.), *Remote Sensing of the Atmosphere: Inversion Methods and Applications*, Elsevier, Amsterdam, 1978.

A collection of papers on inverse problems related to remote sensing of the atmosphere. Topics include temperature sounding, composition sounding and particulate sounding.

[GEZ] J. Gal-Ezar and G. Zwas, Real-world models in the teaching of calculus, *The UMAP Journal* 13(1992), 93-100.

The inverse problem of determining the density distribution with depth of a very simple model of the earth is proposed as an interesting application for the calculus classroom.

[GW] M. Gehatia and D.R. Wiff, Solution of Fujita's equation of equilibrium sedimentation by applying Tikhonov's regularizing functions, *Journal of Polymer Science*, A-2, 8(1970), 2039-2050.

The title says it all.

[Gl] G.M.L. Gladwell, *Inverse Problems in Vibration*, Martinus Nijhoff, Dordrecht, 1986.

A largely self-contained treatment, at a relatively elementary level, of the inverse problem of determining a vibrating system from given spectral properties.

[G] V.B. Glasko, *Inverse Problems of Mathematical Physics*, American Institute of Physics, New York, 1984.

The translation could be better, but this little monograph, by one of the pioneers in the field, gives a down-to-earth account, with some interesting examples, of some now classical work on the theory of regularization of inverse problems.

[God] K. Godfrey, *Compartmental Models and Their Application*, Academic Press, London, 1983.

A good introduction to compartmental modelling and analysis. The main topics are identifiability of linear time-invariant systems, stochastic models, nonlinear models and many applications. There is a separate chapter on Laplace transform identifiability.

[Go] T.I. Goreau, Quantitative effects of sediment mixing on stratigraphy and biochemistry: a signal theory approach, *Nature* 265(1977), 525-526.

A convolution equation of the first kind is suggested as a model for the process of mixing of underwater sediments.

[GV] R. Gorenflo and S. Vessella, *Abel Integral Equations: Analysis and Applications*, LNM 1461, Springer-Verlag, New York, 1991.

A valuable and very readable source on the basic theory, applications, and numerical analysis of Abel integral equations.

[Gro1] C.W. Groetsch, *Generalized Inverses of Linear Operators: Representation and Approximation*, Marcel Dekker, New York, 1977.

An introduction to the Moore-Penrose generalized inverse for bounded linear operators in Hilbert space. The emphasis is on general approximation methods for the Moore-Penrose inverse.

[Gro2] C.W. Groetsch, *The Theory of Tikhonov Regularization for Fredholm Equations of the First Kind*, Pitman, London, 1984.

Lecture notes from a special course on the theory of Tikhonov regularization for compact linear operator equations of the first kind delivered in Kaiserslautern, Germany in 1983. The emphasis is on convergence theory.

[Gro3] C.W. Groetsch, Remarks on some iterative methods for an integral equation in Fourier optics, in "*Transport Theory, Invariant Imbedding and Integral Equations*" (P. Nelson, et al., Eds.), Marcel Dekker, New York, 1989, pp. 313-324.

An operator theoretic treatment of iterative methods for the image reconstruction problem.

[Gro4] C.W. Groetsch, Differentiation of approximately specified functions, *American Mathematical Monthly* 98(1991), 847-850.

A discussion of the optimal order of accuracy (with respect to the size of the perturbation) for difference approximations to the derivative of an approximately specified function.

[Gro5] C.W. Groetsch, Convergence analysis of a regularized degenerate kernel method for Fredholm integral equations of the first kind, *Integral Equations and Operator Theory*, 13(1990), 67-75.

It is shown that under appropriate circumstances the Fredholm integral equation of the second kind with degenerate kernel which results when quadrature is applied to the Euler equation for the Tikhonov functional leads to a regularization method.

[Gro6] C.W. Groetsch, Inverse problems and Torricelli's law, *The College Mathematics Journal*, to appear

This paper discusses three inverse problems in elementary hydrodynamics suggested by Torricelli's law. The presentation is at the level of elementary algebra, calculus and differential equations and the notions of existence, uniqueness and stability of solutions are highlighted.

[Gr] M. Grosser, *The Discovery of Neptune*, Harvard University Press, Cambridge, MA, 1962.

A non-mathematical historical account of the independent efforts of Adams and Leverrier to solve the inverse problem of planetary perturbations, the subsequent discovery of Neptune and the ensuing priority squabble.

[Gru] F.A. Grünbaum, An inverse problem in transport theory: diffuse tomography, in *Invariant Imbedding and Inverse Problems* (J. Corones, et al., Eds), SIAM, Philadelphia, 1992, pp. 209-215.

The author treats the nonlinear algebraic inverse problem of determining transition probabilities (as opposed to simply finding absorption coefficients in standard tomography) from input-output data for a two-dimensional array of pixels.

[Had] J. Hadamard, *Four Lectures on Mathematics*, Columbia University Press, New York, 1915.

A delightful sequence of four "saturday morning" lectures delivered at Columbia University in 1911. Hadamard gives his views on "correctly set" boundary value problems and their relation to physical reality in the first lecture. The second lecture deals with integral equations.

[Had2] J. Hadamard, *Lectures on Cauchy's Problem in Linear Partial Differential Equations*, Yale University Press, New Haven, 1923.

This book contains an early discussion of the notion of an improperly posed partial differential equation.

[**HH**] G. Hämmerlin and K.-H. Hoffmann (Eds.),*Improperly Posed Problems and Their Numerical Treatment*, Birkhäuser, Basel, 1983.

Seventeen papers on inverse and improperly posed problems. Topics include regularization, parameter identification, inverse problems in differential equations and integral equations of the first kind.

[**Hnk**] M. Hanke, Accelerated Landweber iterations for the solution of ill-posed equations, *Numerische Mathematik* 60(1991), 341-373.

A detailed study of the use of linear semi-iterative methods for solving linear ill-posed operator equations of the first kind. In particular, a notion of optimal speed of convergence is investigated for semi-iterative methods, saturation and converse results are proved and a discrepancy principle for iterative methods is given. The results are illustrated in a variety of numerical examples.

[**HnkH**] M. Hanke and P.C. Hansen, Regularization methods for large-scale problems, *Report UNIC-92-04, Technical University of Denmark*, 1992.

A very useful survey of theoretical and computational methods for ill-posed problems focusing on algorithmic aspects of methods for large-scale problems. The authors present as a numerical illustration of various methods the inverse helioseismology problem (determining the internal rotation of the sun from observations of surface oscillations).

[**Hns**] P.C. Hansen, The truncated SVD as a method of regularization, *BIT* 27 (1987), 534-553.

The TSVD is investigated as a regularization method for the ill-conditioned systems which arise from Fredholm integral equations of the first kind. Perturbation bounds are derived and TSVD is compared with Tikhonov regularization.

[**Hns1**] P.C. Hansen,*Regularization Tools: A MATLAB Package for Analysis and Solution of Discrete Ill-posed Problems*, Danish Computing Center for Research and Education, Lyngby, Denmark, 1992, 98 pp.

An exposition of the main issues and methods for discretized ill-posed problems as well as a tutorial and manual for an very useful package of user-friendly MATLAB routines for regularization. An extensive collection of test problems involving Fredholm integral equations of the first kind is included.

[**Hns2**] P.C. Hansen, Numerical tools for analysis and solution of Fredholm integral equations of the first kind, *Inverse Problems* 8(1992), 849-872.

A very nice survey of the state of the art in numerical methods for the solution of linear algebraic systems arising from discretization of Fredholm integral equations of the first kind. The emphasis is on regularization and use of the singular value decomposition.

[Ha] R.J. Hanson, Integral equations of immunology, *Communications of the Association for Computing Machinery* 15(1972), 883-890.

A computational method for constrained solutions to Fredholm integral equations of the first kind is presented. The method is applied to the hapten binding equation of immunology.

[Hal] R.J. Hanson, A numerical method for solving Fredholm integral equations of the first kind using singular values, *SIAM Journal on Numerical Analysis* 8(1971), 616-622.

A numerical method for Fredholm integral equations of the first kind based on discretization by quadrature and collocation and solution of the resulting linear system by TSVD is proposed and tested on a number of equations.

[He] W.L. Hendry, A Volterra integral equation of the first kind, *Journal of Mathematical Analysis and Applications* 54(1976), 266-278.

A pair of Volterra integral equations of the first kind is used to model the relationship between the density distribution of plutonium atoms in doped zirconium oxide spheres and the emission rate of α-particles.

[Her] G.T. Herman, *Image Reconstruction from Projections: the Fundamentals of Computerized Tomography*, Academic Press, New York, 1980.

A handbook on numerical methods, mainly iterative, for recontruction of objects from projections. The main source of applications is medical imaging.

[H] B. Hofmann, *Regularization for Applied Inverse and Ill-posed Problems*, Teubner, Leipzig, 1986.

Advanced notes with an extensive bibliography. Topics include mathematical modeling and optimization, regularization of deterministic discrete inverse problems, regularization of stochastic discrete inverse problems, and numerical approaches to nonlinear inverse problems.

[HB] J.N. Holt and A.J. Bracken, First-kind Fredholm integral equation of liver kinetics: numerical solutions by constrained least squares, *Mathematical Biosciences* 51(1980), 11-24.

A Fredholm integral equation of the first kind relating the concentration of a substrate in input blood to the liver to the concentration of substrate in the output blood is studied.

[Ho] J. Honerkamp,Ill-posed problems in rheology, *Rheologica Acta* 28(1989), 363-371.

The method of regularization and the maximum entropy method are applied to a linear integral equation of the first kind relating stress and shear rate in a polymeric fluid.

[HTR] J.T. Houghton, F.W. Taylor and C.D. Rodgers, *Remote Sounding of Atmospheres*, Cambridge University Press, Cambridge, 1984.

A useful account of the technology, and some of the mathematics, of remote sensing of atmospheric profiles.

[I] V. Isakov, *Inverse Source Problems*, American Mathematical Society, Providence, 1990.

An advanced research monograph on existence and uniqueness theory for inverse source problems in potential theory.

[Iv] V.K. Ivanov, Integral equations of the first kind and an approximate solution for the inverse problem of potential, *Soviet Mathematics Doklady* 3(1962), 210-212.

A discussion of a general approximate method for nonlinear integral equations of the first kind with an application to the problem of determining the shape of a homogeneous planar mass which produces a given gravitational potential.

[JR] W. Jeffrey and R. Rosner, On strategies for inverting remote sensing data, *The Astrophysical Journal* 310(1986), 463-472.

Various methods for Fredholm integral equations of the first kind, including iterative techniques, Tikhonov regularization, the Backus-Gilbert method, and the maximum entropy method are surveyed. Illustrative examples are drawn from astronomy.

[Jo] F. Joachimstahl, Über ein Attractionsproblem, *Journal fuer die reine und angewandte Mathematik* 58(1861), 135-137.

A very early paper on an inverse problem modeled by an integral equation of the first kind. Given a uniform distribution of mass on a line and the total attractive force as a function of the distance from the line, the unknown law of attraction is sought.

[J] B.F. Jones, Jr., Various methods for finding unknown coefficients in parabolic differential equations, *Communications on Pure and Applied Mathematics* 16(1963), 33-49.

The problem of determining a diffusion coefficient, which depends only on time, in the one-dimensional heat equation is considered. The problem is reduced to the study of a certain nonlinear integral equation.

[Kac] M. Kac, Can one hear the shape of a drum?, *American Mathematical Monthly* 73(Part II)(1966), 1-23.

A classic of mathematical exposition. This paper was also the subject of a film, starring Kac, which unfortunately no longer exists. In the paper Kac gives a theoretical presentation of the inverse problem of determining the shape of a membrane fixed along its boundary (a plane curve) from knowledge of the frequencies of vibration. Recently it has been shown that the answer to Kac's question is "No."

[KN] W.J. Kammerer and M.Z. Nashed, On the convergence of the conjugate gradient method for singular linear operator equations, *SIAM Journal of Numerical Analysis* 9(1972), 165-181.

A convergence proof, with convergence rate estimate, is given for the conjugate gradient method for approximating the minimum norm least squares solution of a linear operator equation of the first kind in Hilbert space.

[Klr] J.B. Keller, Inverse problems, *American Mathematical Monthly* 83(1976), 107-118.

An engaging expository article concentrating on some inverse problems in potential theory,e.g., determining a potential from particle transit times, or determining a scattering potential from the differential scattering cross-section.

[Ke] O.D. Kellogg, *Foundations of Potential Theory*, Springer-Verlag, Berlin, 1929.

This classic book on potential theory is still an excellent source of basic information and examples. The treatment is physically motivated and the only prerequisite is a good course on calculus.

[Kg] J.T. King, A minimal error conjugate gradient method for ill-posed problems, *Journal of Optimization Theory and Applications*, 60(1989), 297-304.

The usual conjugate gradient method for a first kind operator equation is an iterative method in which at each step the residual is minimized over a certain Krylov subspace. In this paper a modified version of the conjugate gradient method is developed which minimizes the error, rather than the residual, at each step.

[Kg1] J.T. King, Multilevel iterative methods for ill-posed problems, in *Ill-posed Problems in the Natural Sciences* (A.S. Leonov,et al., Eds.), VSP Scientific Publishers, Utrecht, 1992.

A general theory of multilevel operators as preconditioners for iterative methods for ill-posed linear operator equations is developed. Bounds for error reduction factors are derived and the ideas are illustrated numerically on integral equations of the first kind arising in inverse heat conduction and image restoration.

[KS] M. Klaus and R. Smith, A Hilbert space approach to maximum entropy reconstruction, *Mathematical Methods in Applied Science* 10(1988), 397-406.

The authors show that the maximum entropy regularization method for certain Fredholm integral equations of the first kind is stable. They also show that or certain reconstruction problems, the entropy-regularized approximations are piecewise constant.

[K] R. Kress, *Linear Integral Equations*, Springer-Verlag, New York, 1989.

A thorough, rigorous and modern treatment of linear integral equations including related material on inverse and ill-posed problems and numerical methods.

[L] D.S. Landis, *Revolution in Time*, Belknap Press of Harvard University, Cambridge, MA, 1983.

A facinating historical account of the quest to measure time. Huygens is a major player; Newton, the Bernoullis and l'Hôpital put in cameo appearances.

[La] L. Landweber, An iteration formula for Fredholm integral equations of the first kind, *American Journal of Mathematics* 73(1951), 615-624.

A landmark paper on an iterative method for Fredholm equations of the first kind which is now often called the "Landweber-Fridman" method.

[Lng] R.E. Langer, An inverse problem in differential equations, *Bulletin of the American Mathematical Society* 39(1933), 814-820.

An early paper on the coefficient determination problem in differential equations. The problem involves the determination of the electrical resistivity of the earth's crust from measurements of the electric potential at the surface.

[LLK] J. Larsen, H. Lund-Andersen, and B. Krogsaa, Transient transport across the blood-retina barrier, *Bulletin of Mathematical Biology* 45(1983), 749-758.

A discussion of a simplified model of an inverse problem related to transport of fluorescein across the retina.

[LL] R. Lattès and J.-L. Lions, *The Method of Quasi-reversibility: Applications to Partial Differential Equations*, American Elsevier, New York, 1969.

The "method" is more in the nature of a general philosophy. The idea is to replace an improperly posed boundary value problem by a nearby properly posed problem. The book is a good source of examples of improperly posed partial differential equations.

[Lav] M.M. Lavrentiev, *Some Improperly Posed Problems of Mathematical Physics*, Springer-Verlag, New York, 1967.

One of the early monographs on the subject. Covers the author's work on analytic continuation, inverse problems in potential theory and the wave equation.

[LRS] M.M. Lavrent'ev, V.G. Romanov and S.P. Shishatskii, *Ill-posed Problems of Mathematical Physics and Analysis*, American Mathematical Society, Providence, 1986.

After two good introductory chapters on models of ill-posed problems and basic concepts related to the theory of ill-posed problems, the book really takes off. Advanced chapters on analytic continuation, boundary value problems for differential equations, Volterrra equations, integral geometry and multidimensional inverse problems fill out the book.

[LRY] M.M. Lavrent'ev, K.G. Reznitskaya and V.G. Yakhno, *One-dimensional Inverse Problems of Mathematical Physics*, American Mathematical Society, Providence, 1986.

An advanced monograph concentrating on inverse source problems for the wave equation.

[LH] C.L. Lawson and R.J. Hanson, *Solving Least Squares Problems*, Prentice-Hall, Englewood Cliffs, 1974.

A well-written, self-contained monograph on theory and numerical methods for least squares problems. The book covers orthogonalization methods, conditioning, generalized inverses, the singular value decomposition and perturbation analysis. Computer codes and some illuminating numerical examples are included.

[Lee] D.A. Lee, On the determination of determination of molecular weight distributions from sedimentation-equilibrium data at a single rotor speed, *Journal of Polymer Science* (A-2) 8(1970), 1039-1056.

A study of Fujita's integral equation of the first kind.

[Leo] A.S. Leonov, et al. (Eds.), *Ill-posed Problems in Natural Sciences*, VSP Science Publishers, Utrecht, 1992.

The proceedings of a research conference held in Moscow in August, 1991, during the second Russian revolution. The sixty four papers fall into three categories: theory and methods of solving ill-posed problems, inverse problems in mathematical physics, and applications.

[Lz] P. Linz, *Analytical and Numerical Methods for Volterra Equations*, SIAM, Philadelphia, 1985.

A very readable introduction to Volterra integral equations. Includes existence and uniqueness theorems, numerical methods for first and second kind Volterra equations and ill-posed Volterra equations.

[Lo] A.K. Louis, *Inverse und schlecht gestellte Probleme*, Teubner, Stuttgart, 1989.

Functional analysis based lectures on the theory of ill-posed problems suitable for a graduate seminar. Applications to tomography.

[Lo1] A.K. Louis, Convergence of the conjugate gradient method for compact operators, in [EG], pp.177-183.

The order of convergence of the conjugate gradient method for a compact linear operator equations of the first kind is derived based on certain smoothness assumptions on the data.

[**Lu**] D.G. Luenberger, *Linear and Nonlinear Programming,* Second Edition, Addison-Wesley, Reading, MA, 1984.

A standard text and reference book on methods and theory for mathematical programming and optimization.

[**Maa**] P. Maaß, Generalized Backus-Gilbert methods, in [S4], pp.440-449.

An abstract Backus-Gilbert method is formulated in Hilbert space and the peakedness condition is interpretted as a general problem of best approximation to the delta function in a Sobolev norm.

[**McBS**] J. MacBain and B. Secrest, Source determination in remote sensing, *SIAM Review* 33(1991), 109-113.

The problem of extracting the temporal component of a source term in the acoustic wave equation from observations of the pressure and its normal derivative on a contour is considered.

[**Mar**] J.T. Marti, An algorithm for computing minimum norm solutions of Fredholm integral equations of the first kind, *SIAM Journal on Numerical Analysis* 15(1978), 1071-1076.

A geometrically motivated algorithm, which is essentially Tikhonov regularization applied on a finite dimensional subspace with a discrepancy-like choice of the regularization parameter, is presented and analyzed.

[**Mar2**] J.T. Marti, *Introduction to Sobolev Spaces and Finite Element Solution of Elliptic Boundary Value Problems*, Academic Press, London, 1986.

A very useful and well-organized introduction to the Sobolev space theory and related mathematics necessary for the study of finite element methods for numerical solution of elliptic problems.

[**Mar3**] J.T. Marti, On the convergence of the discrete ART algorithm for the reconstruction of digital pictures from their projections, *Computing* 21(1979), 105-111.

The author gives some relatively simple proofs of the convergence of the algebraic reconstruction technique based on nonexpansive mapping ideas and linear algebra techniques.

[**Ma**] W.V. Mayneord, The distribution of radiation around simple radioactive sources, *British Journal of Radiology* 5(1932), 677-716.

An early paper on the radiotherapy problem. The direct problem of determining dosage curves (curves of constant intensity) for simple radioactive objects, such as needles and rings, is considered.

[McW] J. McGrath, S. Wineberg, G. Charatis and R. Schroeder, Inversion of first-kind integral equations as a plasma diagnostic, in "*Transport Theory, Invariant Imbedding and Integral Equations,*" (P. Nelson, et al., Eds.), Marcel Dekker, New York, 1989.

The problem of calculating radiation density from an x-ray image recorded on film is modeled by an Abel integral equation.

[McI] M. McIver, An inverse problem in electromagnetic crack detection, *IMA Journal on Applied Mathematics* 47(1991), 127-145.

A Fredholm integral equation of the first kind modeling the lower edge of a crack in a metal sheet is developed.

[Mc] D.W. McLaughlin (Ed.), *Inverse Problems,* SIAM, Philadelphia, 1983.

Papers from a conference held in New York in 1983. The three main areas of coverage are: geophysical inverse problems, tomography and inverse problems, mathematical inverse theory and the maximum entropy method.

[McL] J.R. McLaughlin, Analytical methods for recovering coefficients in differential equations from spectral data, *SIAM Review* 28(1986), 53-72.

A useful review of exact methods for recovering coefficients in second and fourth order linear differential equations from knowledge of the eigenvalues and other spectral data.

[Me] W. Menke, *Geophysical Data Analysis: Discrete Inverse Theory,* Academic Press, Orlando, 1984.

A very accessible book on inverse theory intended for undergraduates in geophysics. From the outset all problems are modeled discretely as matrix equations. Random, as well as deterministic, problems are treated.

[MH] K. R. Meyer and G. R. Hall, *Introduction to Hamiltonian Dynamical Systems and the N-Body problem,* Springer, New York, 1992.

This book has nothing to do with inverse problems.

[Mi] J. Milstein, The inverse problem: estimation of kinetic parameters, in "*Modelling of Chemical Reaction Systems*" (K. Ebert, P. Deuflhard, and W. Jäger, Eds.), Springer-Verlag, Berlin, 1981, pp.92-101.

Least squares methods for approximating coefficients in systems of nonlinear ordinary differential equations are developed. Performance of the methods on several inverse problems in chemical kinetics is reported.

[MIS] S. Miyamoto, S. Ikeda, and Y. Sawaragi, Identification of distributed systems and the theory of regularization, *Journal of Mathematical Analysis and Applications* 63(1978), 77-95.

Tikhonov regularization for linear operator equations of the first kind, in which the solution is constrained to lie in a given closed convex set, is applied to the problem of identifying distributed coefficients and source terms in certain parabolic problems.

[M] V.A. Morozov, *Methods for Solving Incorrectly Posed Problems*, Springer-Verlag, New York, 1984.

A hard-to-read translation from the Russian summarizing the work, through 1974, of the leading school in the abstract theory of ill-posed problems.

[Mu] D.A. Murio, The mollification method and the numerical solution of an inverse heat conduction problem, *SIAM Journal on Scientific and Statistical Computing* 21(1981), 17-34.

The regularization approach to unstable inverse problems is to modify the operator to mitigate the effect of errors in the data. Another approach is to "mollify" the data by smoothing with a convolution operator. In this paper the mollification method is applied to the ill-posed inverse heat conduction problem.

[Mu2] D.A. Murio, Automatic numerical differentiation by discrete mollification, *Computers and Mathematics with Applications* 13(1987), 381-386.

The instability of numerical differentiation is well-known. In this note stable numerical differentiation is achieved by smoothing the function with a "mollifier" and differentiating the result. A procedure is proposed to choose the radius of mollification by a discrepancy-like method.

[N] M.Z. Nashed, Operator theoretic and computational approaches to ill-posed problems with applications to antenna theory, *IEEE Transactions on Antennas and Propagation* AP-29(1981), 220-231.

A lucid survey of general operator theoretic methods (regularization, reproducing kernel Hilbert spaces, iterative methods, projection methods) for generalized inversion of ill-posed linear operator equations of the first kind.

[N1] M.Z. Nashed (Ed.), *Generalized Inverses and Applications*, Academic Press, New York, 1976.

This proceedings of a seminar held in 1973 is full of well-written papers containing a wealth of information on generalized inverses (and operators). Topics include the theory of generalized inverses, generalized inverses in analysis, computational methods and approximation theory, and applications.

[N2] M.Z. Nashed, On nonlinear ill-posed problems I: Classes of operator equations and minimization of functionals, in "Nonlinear Analysis and Applications" (V. Lakshmikantham, Ed.), Dekker, New York, 1987, pp.351-373.

An informative survey, with an large bibliography, of nonlinear ill-posed operator equations and extremal problems.

[N3] M.Z. Nashed, A new approach to classification and regularization of ill-posed operator equations, in [EG]], pp. 53-75.

A study of the role of outer inverses as regularizers of ill-posed linear operator equations in Banach space.

[Na] F. Natterer, *The Mathematics of Computerized Tomography*, Teubner, Stuttgart, 1986.

A very professional treatment of the mathematics of the Radon transform and other transforms related to tomography. The book also contains a chapter on abstract ill-posed problems.

[Na1] F. Natterer, Numerical treatment of ill-posed problems, in [Ta]. pp. 142-167.

A survey of numerical techniques for ill-conditioned linear systems which result from discretization of an ill-posed linear integral equation of the first kind. Topics discussed include generalized and regularized solutions, perturbation theory, the singular value decomposition, iterative methods and applications to tomography.

[Ne] A. Neubauer, Finite-dimensional approximation of constrained Tikhonov regularized solutions of ill-posed linear operator equations, *Mathematics of Computation* 48(1987), 565-583.

A theory of Tikhonov regularization for best approximate solution of ill-posed linear operator equations, in which solutions are constrained to lie in a closed convex set, is developed.

[Ol] D.W. Oldenburg, An introduction to linear inverse theory, *IEEE Transactions in Geoscience and Remote Sensing* GE-22(N0.6)(1984), 665-674.

An overview of some methods in linear inverse theory used by geophysicists, concentrating on the Backus-Gilbert method.

[OS] F. O'Sullivan, A statistical perspective on ill-posed inverse problems, *Statistical Science* 1(1986), 502-518.

An interesting and very readable treatment of ill-posed problems from the statistical prospective. The author shows that successful application of statistical regularization methods requires estimates of the signal-to-noise ratio and methods of obtaining such estimates are discussed. The ideas are illustrated with some inverse problems arising in medicine, meteorology and geophysics.

[P] R.L. Parker, Understanding inverse theory, *Annual Review of Earth and Planetary Science* 5(1977), 35-64.

An exposition of inverse theory from the point of view of a geophysicist. The author discusses linear and nonlinear problems and the paper has an extensive bibliography on the geophysical inverse theory literature.

[Pa] A. Papoulis, *The Fourier Integral and Its Applications*, McGraw-Hill, New York, 1962.

A very readable treatment of the Fourier (and Laplace) transform with engineering applications.

[P] L.E. Payne, *Improperly Posed Problems in Partial Differential Equations*, SIAM, Philadelphia, 1975.

A monograph, with an extensive bibliography, on ill-posed problems in partial differential equations concentrating on existence and continuous dependence results for Cauchy problems.

[PL] V. Petkov and R. Lazarov (Eds.), *Integral Equations and Inverse Problems*, Longman, Harlow, U.K., 1991.

A collection of research papers on integral equations and inverse problems with emphasis on inverse scattering problems.

[Ph] D.L. Phillips, A technique for the numerical solution of certain integral equations of the first kind, *Journal of the Association for Computing Machinery* 9(1962), 84-97.

This paper is historic. It barely predates Tikhonov's paper and is concerned with a method, closely resembling Tikhonov's method, for damping instabilities in approximate solutions for Fredholm integral equations of the first kind. The method of regularization is often called "Tikhonov-Phillips" regularization because of this paper.

[PT] J. Pöschel and E. Trubowitz, *Inverse Spectral Theory*, Academic Press, New York, 1987.

A self-contained and masterful treatment of a single one-dimensional inverse problem: the determination of the coefficient in the Dirichlet problem for a simple second order linear ordinary differential equation from knowledge of the eigenvalues. The emphasis is on existence and uniqueness theory (isospectral sets of coefficients, characterization of eigenvalues), not computation.

[Pr] R.T. Prosser, Can one see the shape of a surface?, *American Mathematical Monthly* 84(1977), 259-270.

Inspired by Kac's article [Kac], the author investigates the extent to which the shape of a closed surface is determined by the asymptotic behavior of the eigenfunctions of the exterior Dirichlet problem for the Laplacian.

[R] A. Ramm, Optimal estimation of limited noisy data, *Journal of Mathematical Analysis and Applications* 125(1987), 258-266.

A discussion of the convergence of the Backus-Gilbert method for the case of error-free data and some optimality properties of the method in the case of noisy data.

[RN] F. Riesz and B. Sz.-Nagy, *Functional Analysis*, Ungar, New York, 1955 (translation of the second French edition of 1952 by L. Boron).

A classic text on linear operators that is as alive today as it was forty years ago.

[Ri] E. Rietsch, The maximum entropy approach to inverse problems, *Journal of Geophysics* 42(1977), 489-506.

A lucid exposition of the maximum entropy method with an illustration of the use of the method to estimate the density distribution of a spherically symmetric earth from measurements of its radius, total mass and moment of inertia.

[Ro] G.F. Roach (Ed.),*Inverse Problems and Imaging*, Longman Scientific, Harlow, U.K., 1991

Proceedings of a research conference held at the University of Strathclyde. Topics include inverse scattering, electrical impedance tomography, inverse moving boundary value problems and inverse problems in plane-wave detection.

[Rob] J.A. Roberts (Ed.), *Indirect Imaging*, Cambridge University Press, New York, 1984.

A collection of papers on remote sensing, primarily concerned with problems in radio astronomy. The book includes a chapter of six papers on the maximum entropy method.

[RW] S. Rubinow and A. Winzer, Compartmental analysis: an inverse problem, *Mathematical Biosciences* 11(1971), 203-247.

A useful review paper on compartmental analysis.

[Ru] W. Rundell, Determination of an unknown non-homogeneous term in a linear partial differential equation form overspecified boundary data, *Applicable Analysis* 10(1980), 231-242.

The problem of determining a forcing term in a parabolic equation, which is a product of a function of space with a function of time, given one of these functions and overspecified boundary data, is investigated.

[RB] B. Rust and W. Burrus, *Mathematical Programming and the Numerical Solution of Linear Equations*, American Elsevier, New York, 1972.

A nice introduction to numerical methods based on mathematical programming techniques for the numerical solution of ill-conditioned linear systems arising from Fredholm integral equations of the first kind.

[S1] P. Sabatier, *Some Topics on Inverse Problems*, World Scientific, Singapore, 1988.

The proceedings of an interdisciplinary workshop on inverse problems held in Montpellier, France in 1987. The twenty six papers emphasize inverse problems in physics, particularly inverse scattering theory and impedance tomography.

[S2] P. Sabatier (Ed.), *Applied Inverse Problems*, Springer-Verlag, Berlin, 1978.

Proceedings of a physics conference on inverse problems. The topics of the individual papers include geophysical inverse problems, inverse scattering problems, problems in optics, and nonlinear inverse problems.

[S3] P. Sabatier (Ed.), *Basic Methods of Tomography and Inverse Problems*, Adam Hilger, Bristol, 1987.

Expository lectures on tomography and related inverse problems by G.T. Herman, H.K. Tuy, K.J. Langenberg and P.C. Sabatier.

[S4] P. Sabatier (Ed.), *Inverse Methods in Action*, Springer-Verlag, Berlin, 1990.

Seventy four papers from a conference on inverse problems held in Montpelier, France in 1989. Topics covered include tomographic inverse problems, identification of distributed parameters, spectral inverse problems, inverse scattering, imaging, and nonlinear inverse problems.

[Sa] R.L. Sandland, Mathematics and the growth of organisms - some historical impressions, *The Mathematical Scientist* 8(1983), 11-30.

A excellent exposition of the development of the theory of mathematical growth laws.

[SPSH] F. Santosa, Y.-H. Pao, W. Symmes and C. Holland (Eds.), *Inverse Problems in Acoustic and Elastic Waves*, SIAM, Philadelphia, 1984.

Proceedings of an international conference on inverse problems for acoustic and elastic waves held at Cornell University in 1984. The volume contains twenty three papers with emphasis on inverse scattering, inverse problems in geophysics, mechanics and ocean acoustics.

[STV] D.H. Sattinger, C. Tracey and S. Venakides (Eds.), *Inverse Scattering and Applications*, Contemporary Mathematics, vol. 122, American Mathematical Society, Providence, 1991.

Thirteen papers from an AMS-IMS-SIAM Joint Summer Research Conference held at the University of Massachusetts in 1990. Topics include inverse scattering problems, inverse problems in higher dimensions, inverse conductivity problems, and numerical methods. The volume contains a particularly useful survey, by Cheney and Isaacson, of inversion algorithms for impedance imaging.

[SB] B. Schomburg and G. Berendt, On the convergence of the Backus-Gilbert algorithm, *Inverse Problems* 3(1987), 341-346.

The convergence of the Backus-Gilbert method, with error-free data, is investigated under the assumption that the measurement functionals are linearly independent and complete in an appropriate Hilbert space.

[SP] K.P. Singh and B. Paul, A method for solving ill-posed integral equations of the first kind, *Computational Methods in Applied Mechanics and Engineering* 2(1973), 339-348.

The method of regularization is applied to obtain a numerical solution to a Fredholm integral equation of the first kind which models the contact pressure between two elastic surfaces.

[Sq] W. Squire, A simple integral method for system identification, *Mathematical Biosciences* 10(1971), 145-148.

A method for estimating (constant) parameters in linear differential equations by integration by parts and least squares fitting is described.

[St] H. Stark (Ed.), *Image Recovery: Theory and Application*, Academic Press, New York, 1987.

A collection of expository papers on various mathematical and engineering aspects of the image recovery problem. Of particular interest to mathematicians are the papers by Rushforth on functional analysis and Fredholm integral equations of the first kind and Youla on convex projections.

[Str] J.W. Strutt (Baron Rayleigh), *The Theory of Sound*, two volumes bound in one, Dover, New York, 1945 (reprint of the 1877 edition published by Macmillan).

The pioneering work in acoustics. In it (p. 217) Lord Rayleigh suggests the inverse problem of determining the density of a string from its movements.

[SJ] X. Sun and D.L. Jaggard, The inverse black body radiation problem: a regularized solution, *Journal of Applied Physics* 62(No.11)(1987), 4382-4386.

Tikhonov regularization is applied to solve the Fredholm integral equation of the first kind for the inverse black body radiation problem.

[Ta] G. Talenti (Ed.), *Inverse Problems*, LNM 1225, Springer-Verlag, New York, 1986.

A series of lectures given at a summer school on inverse problems held at Monte Catini Terme, Italy. Topics include inverse eigenvalue problems, regularization methods, tomography, numerical methods for ill-posed problems, and integral equations arising in optics.

[Tar] A. Tarantola, *Inverse Problem Theory*, Elsevier, Amsterdam, 1987.

This is both a textbook and a reference manual on methods for inverse problems. The first part of the book deals exclusively with discrete problems; continuous problems are taken up in part two. The applications are drawn mainly from geophysics.

[**Ti**] A.N. Tihonov (Tikhonov), Solution of incorrectly formulated problems and the regularization method, *Soviet Mathematics Doklady* 4(1963), 1035-1038.

The original paper that started an avalanche of activity in the theory of regularization. See also [Ph].

[**TA**] A.N. Tikhonov and V.Y. Arsenin, *Solutions of Ill-posed Problems*, Winston and Sons, Washington, 1977.

The first book in English on the general subject of ill-posed problems. It contains a number of annoying misprints and the translation is not the best, but it is still an important primary source.

[**TG**] A.N. Tikhonov and A.V. Goncharsky (Eds.), *Ill-posed Problems in the Natural Sciences*, MIR, Moscow, 1989.

A collection of papers on ill-posed problems in the natural sciences (geophysics, electrodynamics, seismology, optics and image processing) modeled by integral equations of the first kind. The emphasis is on the use of regularization methods to obtain numerical solutions.

[**T**] S. Towmey, *Introduction to the Mathematics of Inversion in Remote Sensing and Indirect Measurement*, Elsevier, Amsterdam, 1977.

Twomey, an atmospheric physicist, is one of the pioneers in the theory of regularization for Fredholm equation of the first kind. The introduction contains a number of examples (mainly hard to understand for nonphysicists) of integral equations in atmospheric physics and other remote sensing situations. Linear algebra techniques are emphasized.

[**TR**] I.R. Triay and R.S. Rundberg, Determination of selectivity coefficient distributions by deconvolution of ion-exchange isotherms, *Journal of Physical Chemistry* 91(1987), 5269-5274.

The method of regularization is applied to a Fredholm integral equation of the first kind modeling the probability density function of the selectivity coefficient in an ion-exchange reaction.

[**Tr**] F.G. Tricomi, *Integral Equations*, Interscience, New York, 1957.

A classical (i.e., non-functional analytic) treatment of integral equations.

[**Tru**] M.R. Trummer, Reconstructing pictures from projections: on the convergence of the ART algorithm with relaxation, *Computing* 26(1981), 189-195.

A proof of the convergence of the algebraic reconstruction technique, with relaxation parameters to improve convergence, is given.

[**V**] J.M. Varah, A practical examination of some numerical methods for linear discrete ill-posed problems, *SIAM Review* 21(1979), 100-111.

A study of numerical methods for the ill-conditioned linear algebraic systems which arise from the discretization of a Fredholm integral equation of the first kind.

[V1] J. Varah, Pitfalls in the numerical solution of linear ill-posed problems, *SIAM Journal of Scientific and Statistical Computing* 4(1983), 164-176.

The numerical difficulties associated with ill-conditioned linear systems obtained from discretized Fredholm equations of the first kind are pointed out and the SVD method and the method of regularization are reviewed. Numerical examples involving inverse Laplace transforms are provided.

[Va] V.V. Vasin, The stable evaluation of a derivative in the space $C(-\infty,\infty)$, *USSR Computational Mathematics and Mathematical Physics* 13(1973), 16-24.

It is shown that the process of numerical differentiation may be stabilized by convolving the approximate function with the exact derivative of a certain mollifying kernel.

[Ve] S. Vessella, Locations and strengths of point sources: stability estimates, *Inverse Problems* 8(1992), 991-917.

The problem of determining the locations and strengths of a (known) finite number of point sources in three-dimensional space from measurements of the potentials that the point sources generate is considered. The author shows that, relative to natural metrics, the locations and strengths are Lipschitz continuous with respect to the generated potentials.

[Vo] C.R. Vogel, Optimal choice of truncation level for the truncated SVD solution of linear first kind integral equations when the data are noisy, *SIAM Journal on Numerical Analysis* 23(1986), 109-117.

The truncated SVD method, with the level of truncation chosen by the generalized cross validation method, is investigated when the data consists of discrete values contaminated by white noise error. Convergence rates for the expected value of the square error are obtained under certain assumptions on the decay rates of the singular values.

[Vo1] C.R. Vogel, An overview of numerical methods for nonlinear ill-posed problems, in [EG], pp. 231-245.

A good survey of some optimization techniques for nonlinear ill-posed problems, including the Levenberg-Marquart method, penalized least squares and constrained least squares.

[Wa] G. Wahba, *Spline Models for Observational Data*, SIAM, Philadelphia, 1990.

A Fredholm integral of the first kind becomes, in the limiting case of a kernel which is a linear combination of delta functions, a problem of interpolation. If the interpolation problem is approached in a least squares sense, with additional smoothing imposed by a regularization term, one arrives at the concept of a smoothing spline. These issues and related statistical topics are discussed in this monograph.

[Wal] G. Wahba, Practical approximate solutions to linear operator equations when the data are noisy, *SIAM Journal on Numerical Analysis* 14(1977), 651-667.

An important method for choosing the regularization parameter in Tikhonov regularization, the method of generalized cross validation, is introduced in this paper. The method is statistically based and relies on the actual values of the data, rather than the overall error level in the data.

[We] J. Weertman, Relationship between displacements on a free surface and the stress on a fault, *Bulletin of the Seismological Society of America* 55(1965), 946-953.

The problem of the title is modeled as a Fredholm integral equation of the first kind. It is essentially the same equation as that which models the gravitational edge effect (see [D]).

[W] S.D. Wicksell, The corpuscle problem, *Biometrica* 17(1925), 84-99.

A discussion of the stellar stereography problem for globular clusters and related problems in biology.

[Wil] J. Williams, Approximation and parameter estimation in ordinary differential equations, in "*Algorithms for Approximation*" (J. Mason and M. Cox, Eds.) Chapman and Hall, London, 1990, pp. 395-402.

The problem of determining unknown parameters in systems of ordinary differential equations by nonlinear least squares techniques is discussed. The emphasis is on the conditioning of the least squares problem with respect to the unknown parameters.

[Wi] G.M. Wing, *A Primer on Integral Equations of the First Kind: The Problem of Deconvolution and Unfolding*, SIAM, Philadelphia, 1992.

An excellent introduction to linear integral equations of the first kind arising in inverse problems. The book assumes relatively little in the way of prerequisites and the style is informal and engaging.

[Y] M. Yamaguti, et al. (Eds.), *Inverse Problems in Engineering Sciences*, Springer-Verlag, Tokyo, 1991.

Research and survey papers from a satellite conference of the International Congress of Mathematicians held in Osaka in 1990. Topics include regularization theory, inverse scattering, inverse problems in synthesis and optimization, and mathematical theory of inverse problems.

[**Yo**] D.C. Youla, Mathematical theory of image restoration by the method of convex projections, in [St], pp.29-78.

A survey, using methods of the theory of nonexpansive mappings in Hilbert space, of ART-type reconstruction algorithms applied to general projections on closed convex sets.

[**ZS**] B.N. Zakhariev and A.A. Suzko, *Direct and Inverse Problems: Potentials in Quantum Scattering*, Springer-Verlag, Berlin, 1990.

A well-written introduction to the direct and inverse theory relating scattering data and potential in the one-dimensional Schrödinger equation. The authors take the reasonable approach of discretizing the equation and presenting the direct and inverse problems in the context of algebraic equations.

[**Zhu**] J. Zhu, Using a hypercube to solve inverse problems in reservoir simulations, *SIAM News*, vol. 25, No.2, 1992.

A technical news article about the use of history matching to identify an unknown permeability coefficient in a working oil field.

[**ZR**] K. Zhou and C.K. Rushforth, Image restoration using multigrid methods, *Applied Optics* 30(1991), 2906-2912.

When fine discretization is applied to an ill-posed Fredholm integral equation of the first kind, a large ill-conditioned linear system results. An SVD analysis of such a system is computationally expensive, while standard iterative methods applied to such systems may converge very slowly. In this paper the authors exploit a two level scheme in which an approximation on the fine grid is accomplished by inexpensive iterations on a courser grid. (See also [Kg1])

[**Z**] D. Zidarov, *Inverse Gravimetric Problem in Geoprospecting and Geodesy*, Elsevier, Amsterdam, 1990.

In addition to the inverse gravimetric potential problem, this book also treats the inverse gravimetric problem in geodesy, i.e., the determination of the earth's shape (the *geoid*) from gravity measurements. It also contains a good discussion of gravi-equivalent bodies, that is, distinct bodies producing the same exterior potential in a given region in space.

[**ZN**] V.E. Zuev and I.E. Naats, *Inverse Problems of Lidar Sensing of the Atmosphere*, Springer-Verlag, New York, 1982.

Lidar (laser radar) uses lasers in radar fashion to probe the atmosphere. The idea is to profile particulate concentrations or optical parameters by observing a backscattered laser beam as a function of range.

Index

Lectures on Nonlinear Evolution Equations

Initial Value Problems

by Reinhard Racke

1992. viii, 259 pages (Aspects of Mathematics, Volume E19; edited by Diederich, Klas) Hardcover
ISBN 3-528-06421-8

Serves as an elementary, self-contained introduction into some important aspects of the theory of global solutions to initial value problems for nonlinear evolution equations. The presentation is made using the classical method of continuation of local solutions with the help of a priori estimates obtained for small data.

The existence and uniqueness of small, smooth solutions which are defined for all values of the time parameter is investigated. Moreover, the asymptotic behaviour of the solutions is described as time tends to infinity. Here, the admissible nonlinearities are pertubations of the linearized equations which are small for small values of their parameters.

Vieweg Publishing · P.O. Box 58 29 · D-6200 Wiesbaden 1

vieweg

SPRINGER NATURE

GPSR Compliance

The European Union's (EU) General Product Safety Regulation (GPSR) is a set of rules that requires consumer products to be safe and our obligations to ensure this.

If you have any concerns about our products, you can contact us on ProductSafety@springernature.com

In case Publisher is established outside the EU, the EU authorized representative is:

Springer Nature Customer Service Center GmbH
Europaplatz 3
69115 Heidelberg, Germany

Zeitfracht Medien GmbH
Ferdinand-Jühlke-Straße 7
99095 Erfurt, Deutschland
produktsicherheit@kolibri360.de